THE
GUT-BRAIN
PARADOX

ALSO BY STEVEN R. GUNDRY, MD

The Energy Paradox

Unlocking the Keto Code

The Longevity Paradox

The Plant Paradox

The Plant Paradox Cookbook

The Plant Paradox Family Cookbook

The Plant Paradox Quick and Easy

Gut Check

Dr. Gundry's Diet Evolution

THE GUT-BRAIN PARADOX

Improve Your Mood, Clear Brain Fog, and Reverse Disease by Healing Your Microbiome

Steven R. Gundry, MD

With Jodi Lipper

HARPER

An Imprint of HarperCollins*Publishers*

THE GUT-BRAIN PARADOX. Copyright © 2025 by Steven R. Gundry, MD. All rights reserved. Printed in the United States of America. No part of this book may be used or reproduced in any manner whatsoever without written permission except in the case of brief quotations embodied in critical articles and reviews. For information, address HarperCollins Publishers, 195 Broadway, New York, NY 10007.

HarperCollins books may be purchased for educational, business, or sales promotional use. For information, please email the Special Markets Department at SPsales@harpercollins.com.

FIRST EDITION

Designed by Nancy Singer

Library of Congress Cataloging-in-Publication Data has been applied for.

ISBN 978-0-06-291180-3

25 26 27 28 29 LBC 5 4 3 2 1

To Hippocrates, who famously said, "All disease begins in the gut."

To Ignaz Philipp Semmelweis, MD (1818–1865), for questioning his medical school professors at the Vienna General Hospital as to the real cause of childbed (puerperal) fever, at that time the leading cause of maternal death in childbirth. Prior to the discovery of bacteria, he correctly identified the cause as the lack of hand sterilization between cases, which he instituted for his own cases with profound results. But his practice was soundly repudiated by his peers.

For his insights, he was declared a quack, deported to Budapest, placed into an asylum, and beaten to death.

Fifteen years later, with the discovery of bacteria, his insights were proven and his place as the father of antisepsis heralded around the world! May he rest in peace.

Science is based on experiment, on a willingness to challenge old dogma, on an openness to see the universe as it really is. Accordingly, science sometime requires courage—at the very least, the courage to question the conventional wisdom.

—Carl Sagan

CONTENTS

THE GUT-BRAIN PARADOX

LET FOOD BE THY MEDICINE

About a year ago, a sixty-two-year-old woman who had been suffering from Parkinson's disease for five years with no relief found her way to my clinic. By that point, she was desperate for help. Before developing symptoms of Parkinson's, she was in excellent shape—a practiced marathon runner—but she now suffered from the severe tremors that are a hallmark of the disease.

The medications this woman had been given by her neurologists hadn't helped, and her symptoms kept getting worse. By the time she came to see me, her balance was so off that she was barely able to walk. She remained seated throughout our appointment, with a flat affect and an expressionless face, also known as the "mask of Parkinson's." With her speech halting, the patient's husband did most of the talking.

I ran my usual blood tests and put this patient on a food and supplement program based on my findings. My physician's assistant did a follow-up phone consultation six months later, and the patient was seeing some improvements. Six months after that, a year after I first met her, I walked back into the same examination room I had first seen this patient in.

She was a completely different woman. At first, I thought it was literally a different patient. I assumed I had mistakenly entered the wrong room. But it was her. She had a big smile on her face and zero tremors. Her husband, who was sitting next to her, jumped up as soon as he saw me and gave me a huge hug.

"Thank you for giving me my wife back," he said, his eyes misting. "This is the girl I married. She is back to running every day—I can't keep up with her!"

Soon, all three of us were crying and hugging, with the grateful couple thanking me profusely. But I assured them: I'm no miracle worker. I provided an eating and supplement protocol—the rest of the healing work, she did herself. And my amazing patients do the same thing every single day.

Recently, another patient of mine saw an equally remarkable turnaround, this time a high schooler. She had been an impressive athlete before developing a movement disorder and postural ortho-static tachycardia syndrome (POTS), which causes patients to get dizzy when standing up. When she came to see me, this young woman could only get around with the use of a walker. She had also been diagnosed with bipolar disorder and obsessive-compulsive dis-order. She was a teenager and could no longer attend public school; she was really struggling physically and emotionally.

Like my patient with Parkinson's, based on her blood test results, I put this young woman on my protocol. Within a month, she was walking with a cane, and her erratic behavior had normalized. After a year, she had ditched the cane, had returned to her regular school, and was even playing on several sports teams. The change in her was dra-matic. Sometimes, she still struggles with mild obsessive-compulsive behaviors, but even those are much improved. Most recently, she was on a major carrot kick—eating a large amount of carrots every day. I told her that this type of fixation was fine with me!

I also saw a young man in his thirties who had struggled with addiction to drugs and alcohol for decades. He had been in and out of literally a dozen different rehabilitation centers and treatment pro-grams over the course of many years, but nothing had worked. He was struggling with his mental health and unable to hold a job. Of course, his family was beside themselves with worry that he would become another victim of addiction. But, like my other patients, he was able to turn his health and life around by following my protocol. He is now celebrating several years of sobriety.

What if I told you that I put these three patients on the exact same protocol, and it's the one you are going to learn in this book? Would you believe that one protocol consisting simply of foods and supplements could successfully treat such vastly different conditions?

What if I also told you that this is not only true, but that these three patients were actually suffering from the exact same condition? It was simply manifesting in the form of different symptoms in each individual. My tests revealed that these patients (and nearly all of those whom I treat) were suffering from intestinal permeability (aka "leaky gut") and gut dysbiosis, a disruption to the microbiome leading to an imbalance of microbiota. This is what caused their symptoms. And if you are suffering from brain fog, mental fatigue, neurodegeneration, mental health issues, addiction, or other behavioral or cognitive problems, it is most likely being caused by the same underlying problem.

How is this possible? There is a vast, incredibly complex communication system between the microbes in your gut and your brain. It may be hard to believe at first, but all rapidly accumulating evidence points to the fact that your microbes are the ones in charge.

We'll explore this system in great detail throughout this book, but in short, microbes send signals to your brain telling it (you) what to think, how to feel, and even how to behave. Through these systems, our microbiomes drive brain-related illnesses and even help shape our personalities and our emotional states.[1] As you'll learn, when your gut biome becomes imbalanced and pathogenic "bad bugs" take over, they hijack these communication systems and begin calling the shots, and you and your brain pay the price.

I have argued in my previous books that Hippocrates was right when he said that all disease begins in the gut. Yet, many of the things

I learned while doing research for this book about how our micro-biomes are controlling our brains truly surprised me. Not only was Hippocrates right that all disease begins in the gut, but I would also argue that nearly all mental health, cognitive, behavioral, personality, habit-based, and neurodegenerative issues also begin in the gut. In fact, I will spend the rest of this book arguing that case.

As just a small preview of the type of mind-blowing research I'll share, it turns out that many of the factors influencing human behavior that we previously thought had nothing at all to do with the gut absolutely do. For instance, humans have believed for centuries that the phases of the moon impact behavior, but we've never before connected this to the microbiome. While many have written about these ideas about the moon as mere superstition, I can tell you that back when I was an ER doc, we always staffed heavily on nights when there was a full moon. There is a higher rate of traffic accidents during a full moon.[2] Even patients with bipolar disorder experience mood cycles that mirror the phases of the moon.[3]

However, we never understood the mechanism behind the moon's impact on our behavior. Many chalked it up to the ways that sleep varies with the lunar cycle.[4] Makes sense, right? Tired patients are more likely to have a traffic accident. Perhaps tired patients with bipolar disorder are more likely to have extreme mood swings.

But why exactly does the moon affect our ability to sleep? Lo and behold, our microbiomes have their own circadian rhythms! Certain bacteria are more plentiful at particular times of the day. Of course, bacteria cannot see the moon, but they (and their circadian rhythms), like ocean tides, are affected by its gravitational pull.[5] And as you'll see, changes to the microbiome—even those caused by the moon—directly lead to changes in mood, behavior, mental health, and so much more.

Unfortunately, our Western diets and toxic environments—not to mention our obsession with sterilization and killing off the very bacteria that we need to live—have destroyed our microbiomes. This is why nearly every patient I see has leaky gut and dysbiosis. It's also why we are seeing such a stark uptick in neurodegenerative diseases, mental health issues, addictions, and so on.

The good news is that the mental and neurological problems that begin in the gut can also be halted, reversed, healed, and even prevented in the gut. Just like the patients I mentioned, you will see truly dramatic turnarounds in your own cognition and mental health once you balance your microbiome and begin experiencing what it's like to have the right messages being sent from your gut to your brain. The result is a sharper mind, healthier habits, greater feelings of well-being, and the reversal of neurodegenerative symptoms.

Some of the information in this book may come as a shock. Truthfully, much of it came as a shock to me. It asks us to reconsider who we are as humans and why we act, feel, and think in the ways that we do. But I also believe that there is tremendous freedom and peace that come with knowing that, to a great extent, our brains are under the direction of our microbiomes. We need not feel guilt or shame because of an addiction or a brain alteration. None of these things are our fault.

It's also not our fault that we've unwittingly killed off the microbes that are required for brain health. We have simply been given bad information. As Maya Angelou said, "Do the best you can until you know better. Then, when you know better, do better." I trust that you have been doing the best you could until now. And by the end of this book, you will know better and have the ability to do better.

In this book, we will first explore the fascinating ways in which the microbiome controls the brain. Then, we'll move on to take a close look at how dysbiosis and leaky gut can lead to common brain conditions such as addiction, mental health issues, and neurodegeneration. Finally, I'll provide two food plans to choose from based on your personal goals that will teach you exactly how to eat for a balanced microbiome, and therefore a sharp, energetic, healthy, and thriving brain.

It's time to do better! Let's get started.

CHAPTER 1

SH*T FOR BRAINS

Back in the 1800s, one of the most hotly contested "celebrity" feuds was, believe it or not, between two French chemists: Louis Pasteur and Antoine Béchamp (with his colleague Claude Bernard). Chances are, you've only heard of one of these scientists, which means that he ostensibly "won" the debate. This is true, but it does not tell the full story, or even close.

Pasteur and Béchamp were fighting about germs, also known as microbes or microorganisms. As you probably know, these are tiny living things that are too small to be seen by the naked eye, and they include viruses, bacteria, archaea, fungi, and protists. Both men agreed that these microbes existed, although there was some debate about which one of them discovered microbes and their role in fermentation. What they mainly disagreed about, however, was whether these microbes were objectively good or bad.

Pasteur believed, and it came to be accepted as fact, that microbes were bad. His development of pasteurization, which he is best known for, partially sterilized milk. This process killed off microbes and made milk safe to drink. But Pasteur took this idea a step further, claiming that microbes caused disease and that killing them off was therefore always the cure. This became known as the germ theory of diseases, which is at the root of modern Western medicine.[1]

Since the days of Pasteur, much effort has gone toward killing off germs—aka sterilization—to prevent and cure disease. Just think about that hand sanitizer you likely carry around with you, not to mention the last round of antibiotics your doctor prescribed. As a society, we love to kill germs. Of course, some specific bacteria, such as some *E. coli* and *Salmonella*, are pathogenic and therefore likely better off dead. But Pasteur's germ theory was only partially correct, and his rival Béchamp knew it.

Béchamp's argument was that most germs in and of themselves were not the problem and that they only caused disease when the environmental "terrain" in which they lived was disrupted, making the host susceptible to disease. In other words, microbes could only cause disease when the conditions in the body allowed them to overgrow. Béchamp also believed that sterilization was a) impossible, and b) terribly dangerous. He believed that microbes were an essential part of human beings and all living things.[2] Sadly, Béchamp's argument failed to gain any traction. (Pasteur was a brilliant public speaker, and in 1863 he allegedly gained Napoleon III's undying support when he "proved" that wine went "bad" from bacterial contamination.)

With the advent of the ten-year Human Microbiome Project in 2007, we now know that it was Béchamp, not Pasteur, who was in the right. Every day we are discovering shocking new things about the sheer number of microbes living in and around us, not to mention the countless, complex ways they influence the health and behavior of us, their hosts. The microbes in and on our bodies, which in totality make up our holobiome, live primarily in our guts. This is generally referred to as our microbiome. But many other parts of our bodies contain their own mix of microbes.

These microorganisms are an inextricable part of us. I'd go so far as to say that based on what we now know about our holobiomes, we must reconsider who and what we are as humans. Are we lone creatures or symbiotic communities made up of a combination of human and microbial cells? If you believe you are the former, by the end of this book I hope to change your mind. In fact, I hope to change it in many ways.

THE HOLOBIOME

Let's start by looking at the numbers. We now believe that about half of the cells in our bodies are human and the other half are microbial.[3] Bacteria comprise about 1 to 3 percent of our body mass, making up the largest diffuse organ system in the body, one that is at least as metabolically active as the liver. The gut microbiome alone contains at least a hundred trillion bacteria belonging to at least ten thousand different species, plus an as yet undetermined number of viruses, fungi, and other microbes.

We also have an oral biome with seven hundred species of bacteria and a skin biome with a thousand different species. In fact, we have bacteria in every part of the body that interacts with the outside world. This includes our lungs, breast ducts, vagina, uterus, ureter, and prostate. They each have their own microbiome. Between them, these microbes contain over three million genes, while the human genome contains a mere twenty-three thousand.

Yes, you have far more bacterial than human genes inside of you. If that freaks you out even a little, I promise it's just the tip of the iceberg. There is now evidence that we have bacteria living in parts of our bodies that we have always believed to be sterile—until now. For instance, there are bacteria in healthy human blood,[4] and we've recently discovered that the brain has its own microbiome.

That's right; bacteria have been found in both healthy human and mouse brains. Intriguingly, the bacteria found in the brain belong to species that are common in the gut.[5] This begs the question—do these bacteria normally inhabit our brains, or have they translocated from our guts? More on this to come.

You see, I am not using the term "sh*t for brains" as this chapter's title as an insult. I am being quite literal. Not only do we have bacteria in our brains, but the bacteria in our guts and throughout our bodies are, to a great extent, directing what goes on in our human brains. What you'll learn throughout this book is that these microbes don't just live inside of us, whether it's in our guts or our blood or our brains. To an enormous extent, they control us, playing an outsize role in developing our personalities, including how we think, feel, and act; the foods we prefer; and whether or not we'll suffer from addiction, struggle with our mental health, or fall prey to neurodegeneration. If you think this is an overstatement, I promise you it is not.

Hopefully, you're beginning to realize that our gut buddies are not all bad and that—to quickly bust another myth—they do a lot more than simply help us digest food. And even when it comes to digesting food, certain species of bacteria are highly specialized. For instance, some species break down starches, while others ferment proteins. They then deliver vitamins, minerals, and proteins to exactly where they are needed in your body.

As I'll discuss in great detail, your gut buddies also orchestrate your endocrine (hormonal) system, your nervous system (including your brain), and your adaptive immune system. The gut-brain axis is the bidirectional communication channel between your micro-biome and these three all-important systems, which are intricately connected.[6]

Via the gut-brain axis, your microbiome is the main educator and caregiver of your immune system. When anything enters your body from the outside world, your gut buddies send specific messages telling your immune system either to attack because it is unfamiliar and may be a threat or it is a recognizable threat, or to stand down because it is familiar and has already been deemed safe. But when these signals get crossed, it leads to a host of problems within us, the host (host, get it?).

To put it a bit more philosophically, the microbiome helps our immune system distinguish between what is part of our "self" and "non-self" (aka an invader) and whether or not that "non-self" is a threat and should be attacked. Does that not make the microorganisms that we host within us and are making this call a part of us? But I digress . . .

Unfortunately, Béchamp was mostly written off as a quack in his time, and the germ theory of diseases prevailed. This led us not only to overlook the tremendous importance of our holobiomes, but to treat the microbes inside of us, which I call our gut buddies, with downright hostility. We've deprived them of the foods they need to thrive, and we've directly destroyed them with household products, pesticides, and, of course, antibiotics and sterilization.

This has come at a terrible cost in the form of the multiple epidemics and health crises that we are suffering from today. Much has been written about the various factors contributing to the mental health crisis, the opioid crisis, the stark increase in autism diagnoses, and the rise in dementia, Parkinson's, and Alzheimer's disease. In this book, however, I am focusing on how these are all directly tied to the destruction of our microbiomes. As Béchamp warned, we have disrupted our inner terrain, and our brains are now paying the price. Thanks a lot, Louis.

DYSBIOSIS BY ANY OTHER NAME

So, what exactly did Béchamp mean when he talked about our "terrain"? Let's pause for a second and give "terrain" a different name that you might more easily recognize: "ecology." The *Oxford English Dictionary* defines "ecology" as the branch of biology that deals with the relations of organisms to one another and to their physical surroundings. Sounds like a tropical rainforest, right? One organism is dependent on another, whether it's a plant, an insect, a fungus, a bird, or a mammal. They're all interconnected in that unique environment. This is exactly what Béchamp was describing as our "terrain."

So, what exactly were the ecological conditions in the gut that allegedly made a host susceptible to disease? We now know that a healthy inner terrain that sets the stage for a well-functioning brain exists in a state of homeostasis. In this state, there is a stable equilibrium between interdependent elements—in this case, between different microbial species and other forms of microbes.

The key words here are "stable" and "interdependent."

Stability of the Terrain

In a stable terrain, the exact mix of gut buddies present should remain consistent for a long period of time without much fluctuation.[7] The more stable your microbiome is, the more quickly it will bounce back after a short-term disruption, for instance when you are exposed to a pathogen or when you have to take a course of antibiotics. This is one reason that some people get very sick when they are exposed to a pathogen and others recover quickly. In the case of the latter patient, the microbiome was more stable and was therefore able to bounce back faster.[8]

If—and it's an important if—you have the right mix of gut buddies, they will work hard to help keep your microbiome stable, contributing to a healthy homeostatic terrain. For instance, if one or more bacterial species overgrow, threatening the microbiome's overall stability, bacteria called "keystone species" respond by changing the environment to make it more difficult for the ones that are overgrowing to reproduce.[9]

For example, when a bad bug called *Pseudomonas aeruginosa* starts to overgrow, keystone species release acetic acid. This prohibits the bad bug's growth.[10] Microbiome stability is so important that keystone species of bacteria exist explicitly for this purpose.[11]

Single-celled organisms actually sense, respond, and make decisions. In short, they "think." Now, hold on. Single-celled organisms don't "think" like we do. They don't have brains, but the intelligence they do have is downright amazing! They even know how many others of their kind are needed to complete an action and can count whether or not the right number is present. This is called quorum sensing, which, among other things, keeps their numbers stable.

Interdependence of the Terrain

Your gut buddies work together and communicate with one another constantly in order to effectively do their jobs, all with the goal of keeping you healthy and flourishing. After all, you are your microbiome's home. Your body's inhabitants want you to be healthy. However, because they are interdependent, meaning they depend on one another to get things done, your microbes can do their jobs only if you have the right balance and a diverse mix of inhabitants.

Just as diversity has been proven to benefit groups of humans, as those differences lead to a valuable mix of perspectives, it is essential for your gut buddies, too. In ecology, diversity indexes like Simpson's and Shannon's are used to rate the "health" of a rainforest. So, it's not surprising that a diverse gut biome is directly linked to health and longevity.[12]

As your gut buddies complete their individual tasks to keep you thriving and alive, they rely on one another to get things done, often passing the proverbial baton to another species to complete the next step. A great example of this is what's called "cooperative digestion." This is when the compounds that one type of bacteria produces through its digestion process become the food for another bacteria. As those bacteria digest the compound, they produce yet another compound that becomes food for a third species. Beautiful! But if you are missing one or more of these species, the whole system can start to break down. Indeed, low microbial diversity is associated with brain degeneration and many different types of disease.[13,14,15]

But what about those pathogens that Pasteur was so eager to kill off? Are they a part of this microbial diversity? Yes, they are. It turns out that a healthy, stable, and interdependent microbiome includes some bad guys and even parasites.[16] As Béchamp rightly claimed, it's only if the terrain is disrupted that the bad bugs become a problem.

Although we don't have any real proof, it has often been repeated that on his deathbed back in 1895, Pasteur finally admitted, "*Bernard avait raison, le microbe n'est rien. Le terrain est tout.*" "Bernard was right, the microbe is nothing. The terrain is everything."

I like to believe this anecdote to be true, but it doesn't really matter. By then, it was too late. The germ theory of diseases had spread like a virus (get it?) and been so widely accepted as fact that there was no turning back. And in the years since then, we have created absolutely perfect conditions to destroy our internal terrain, from the foods and substances we ingest to the products we use and the ways in which we treat our external, literal soil.

Nearly two hundred years after Béchamp and Bernard were written off, we are just now beginning to understand the importance and the intricate workings of our inner terrain. We now refer to a disrupted microbial terrain as a case of gut dysbiosis, and it is at the root of neurological, neurodegenerative, psychiatric, and many other types of disease.[17]

Bernard avait raison, indeed.

DEATH BEGINS IN THE GUT

If you have a habit of reading scientific longevity studies, it's likely that you've come across a tiny worm called *C. elegans*. These harmless little guys (and I'm not being sexist—there are no female *C. elegans*, just male and intersex) have been studied extensively for a few good reasons. One, *C. elegans* is a very simple organism with a lifespan of just two to three weeks. Two, they're almost completely translucent, so it's easy to see what's going on in there. And three, they possess many of the same essential biological characteristics as humans.

In particular, *C. elegans* is helpful for studying the aging process because it goes through several distinct phases of life that are easily observable within a short time. And that aging process mirrors our own. In fact, nearly every intervention that has been proven to extend the lifespan of *C. elegans* has been found to be reproducible in animals up to the biological complexity of a rhesus monkey.

Notably, the intestine of *C. elegans* is their largest and arguably most important organ, comprising one-third of its mass, and it has its own microbiome. Humans are similar. While some say that our skin is our largest organ, the truth is that it's got nothing on the wall of our guts, which has the surface area of at least a tennis court, maybe two. This is called the intestinal epithelium, which is a single-celled surface lining of both the small and large intestines. Yes, you read that right: A single layer of cells is all that stands between you and everything you swallow. The same design flaw is repeated in our friend *C. elegans*.

In humans and *C. elegans*, the intestinal epithelium, or cell wall, is comprised of tight junctions between cells that keeps everything in its place—food and bacteria inside the intestines and everything else out.[18] When it comes to the gut wall, I like to say that good fences make good neighbors. This is true both for *C. elegans* and for us.

Thanks to its translucence, we can easily observe that the *C. elegans* aging process is driven by this gut wall. As the worm ages, the gut wall becomes porous, and bacteria are able to leak out of the intestine. This is what we call leaky gut. At this point, the *C. elegans* quickly begins to deteriorate. It starts to eat less and become less active and is eventually immobile. Then it dies.

Sound familiar? This is the same aging process that we experience as humans, just rapidly sped up. Both processes are driven by increasing permeability of the gut wall. This leads to inflammation that causes changes in the brain (and elsewhere in the body), which (among other things) coordinates appetite and movement.

We've already established that Hippocrates was correct that all disease begins in the gut. But the more I learn, the more I am willing to take it a step further and say that death begins in the gut. Once the gut wall starts to break down, it's as if the fortress has been breached. The war is over. And death is imminent. It may be easier to believe this is the case when we can see it happening in *C. elegans* with our own eyes, but with the multiple epidemics and crises that humans are now facing, we can no longer afford to pretend the same thing isn't happening within us, even if we can't observe it quite as easily.

I don't mean to sound morbid—well, maybe a bit—but rather to offer hope. I have helped thousands of patients reverse their leaky guts and restore their brain and bodily health. Leaky gut is reversible. So if death begins in the gut, it's there that it can also be stopped and that health can be restored.

After all the books I've written and all the patients with leaky guts whom I've successfully treated, I'm constantly amazed to learn that many people in the field of medicine still claim that leaky gut does not exist. They write me off as a quack, which I actually take as a great compliment because it puts me in good company. (I'm looking at you, Béchamp.) However, it does nothing to help the countless patients who are suffering.

If nothing else, I want you to walk away from this book believing that leaky gut is real and that it can be fixed. How can I convince you? Here's one piece of evidence: When mice were fed a diet that triggered gut dysbiosis and leaky gut, bacteria from their guts were able to translocate from the gut to the brain. When the mice's diet was changed, their gut biomes became more stable and their leaky guts healed. As a result, there were significantly fewer bacteria in their brains.[19] Are we really so arrogant to believe that the same thing isn't happening in humans?

A ONE-TWO PUNCH OF DYSBIOSIS AND LEAKY GUT

Let's take a moment to look at how dysbiosis and leaky gut exacerbate each other. The truth is that if—another big if—we had a healthy internal terrain, the design of the gut wall would work beautifully. It's the disruption of the terrain that drives leaky gut and the neurological problems that stem from there.

If you have read my book *The Plant Paradox*, you know that one of your gut buddies' jobs is to eat and/or destroy harmful compounds that can injure your gut wall, such as lectins. (More on this later.) They act like the offensive line of a football team protecting the quarterback. But sadly, that Front Four has been decimated and malnourished to the point that they can't show up for work, much less protect you.

Thankfully, there is another line of defense: mucus. The gut epithelium is covered with a layer of mucus, and just inside of that wall lies 60 to 80 percent of your entire immune system, in the form of specialized white blood cells. These are your last two lines of defense against any invaders that can attack the gut wall. First and most simply, that mucus is meant to trap invaders before they can get to the gut wall. If that doesn't work, those white blood cells are there to attack anything entering through your gut wall that appears to be a threat.

When white blood cells detect that something threatening has breached the wall, they release inflammatory hormones called cytokines, causing inflammation. This signals to the rest of your immune system that invaders are on the way and they should be prepared to defend themselves from imminent attack. Again, this system would work great if you had a healthy terrain. It's a good thing for your immune system to be ready to attack in the rare instance that a deadly pathogen has indeed crossed the gut wall. In this case, the acute inflammation resulting from the attack would be worth it.

Without a healthy terrain, however, the entire system breaks down. First of all, you need the right mix of gut buddies to maintain your first lines of defense. They can eat unwanted invaders, thwart invaders' growth with signaling compounds, and maintain a robust mucus layer along the epithelium. I know I probably shouldn't play favorites, but if you've read my other books, you probably already know that I'm partial to one species of bacteria in particular, called *Akkermansia muciniphila*, which means "mucus loving."[20] *Akkermansia* eat the mucus layer along your gut and use it to produce butyrate, one of the most important short-chain fatty acids (SCFAs) in your body.[21,22] Yes, butyrate is another favorite, and for good reason.

In a well-balanced terrain, that butyrate produced by *Akkermansia* triggers the cells that line the gut wall to produce more mucus. (Yes, the more mucus the *Akkermansia* eat, the more you make!) This fortifies your defenses. Butyrate also reduces inflammation and protects the health of your mitochondria, the little power plants within your cells. Healthy mitochondria mean healthy cells, so the more butyrate you have along the epithelium, the stronger and healthier your gut wall will be.[23,24]

Sadly, with a disrupted terrain, you likely don't have enough *Akkermansia*, which means you don't have enough butyrate, which means your mitochondria and the cells lining the epithelium aren't as healthy as they should be. Plus, you don't have enough mucus, which means your second line of defense is gone, and the few *Akkermansia* that you do have are dying of starvation because they need that mucus to grow and multiply. This intensifies the problem and perpetuates the cycle.

Even worse, when we don't eat the foods that many of our gut buddies need to live—namely those containing dietary fiber—they start eating the mucus layer out of desperation. The problem is that these guys can't make butyrate out of that mucus like *Akkermansia* can, so the mucus layer gets eaten away without strengthening the cells or sending a signal to produce more mucus. This erodes the mucus layer along with your defenses.[25] Plus, with *Akkermansia* and these other species now competing for mucus as a food source, *Akkermansia* start to die off even more, leading to an even worse case of dysbiosis.

Now you have leaky gut, but what exactly is leaking through? Primarily, it's four things: The first is pathogenic bacteria, which are actual threats. That's no good. Second is lectins, which I mentioned earlier and will be familiar to you if you've read my other books.

Lectins are a type of protein found in many plants that actually evolved as a defense mechanism to keep the plant or its seeds from being eaten by predators.

If you have a healthy terrain (yes, I might be repeating that phrase a lot), lectins shouldn't really be much of a threat. A healthy, robust microbiome normally loves eating lectins and can ferment them into valuable compounds.[26] Barring that, lectins, being proteins, are attracted to mucus, so the mucus layer on your endothelium should trap and bind them.

With a disrupted terrain, however, there's no mucus to bind lectins and no robust microbiome to eat them. So, lectins are able to bind with receptors along the gut lining, instead, and produce a compound called zonulin, which breaks the tight junctions that hold your gut lining together. This makes the gut wall even more penetrable. Definitely no good.

The third type of potential invader is lipopolysaccharides (LPSs), which are fragments of cell walls from dead bacteria. Although I don't normally swear, to help you remember these guys, I like to say that LPS stands for "little pieces of sh*t," because that is exactly what they are! I've known for a long time that LPSs are a major source of immune system activation and therefore inflammation. However, as I'll discuss later, I've also recently discovered that the correct amount of the right LPSs introduced into your diet can actually train your immune system to get comfortable with LPSs, cutting down on inflammation!

Finally, if the wall of the gut has gaps, undigested food particles that would normally never make it across the gut wall without first being broken down and absorbed can leak through. These appear to our immune system to be foreign invaders (literally splinters) and are attacked, as well.

You may already be wondering why the immune system is attacking the food particles and innocent pieces of dead bacteria that leak through a permeable gut. The simple answer is that without the right mix of gut buddies to educate the immune system in the first place, our white blood cells literally don't know how to recognize a threat. So, they take the approach of "better safe than sorry," sounding the alarm nearly constantly, to the detriment of your brain.

A POORLY TRAINED MILITIA LEADS TO
NEUROINFLAMMATION

Let's back up a second and look at how immune cells in the gut are supposed to be able to differentiate between an invader and a gut buddy. As you might imagine, your immune system is designed primarily to protect you from harmful bacteria, viruses, and other unwanted microbes. Your white blood cells contain sophisticated scanning devices called toll-like receptors (TLRs) that recognize the structure of specific molecules.

As a molecule comes across the gut wall, the TLRs scan it to determine if it's foreign and a threat or if it's familiar and known to be nothing to worry about. Again, this system should work great, but when you have the one-two punch of leaky gut and dysbiosis, it doesn't. For instance, in mice, leaky gut leads to alterations in TLR signaling, gut microbial composition, and significant cognitive impairment.[27]

Your immune system "reads" the barcode on LPSs as threatening living bacteria, even though they're dead. And your immune system then triggers inflammation to deal with the threat.[28] Why does this happen? Since the days of Pasteur, we've been so intent on killing bacteria and preventing our immune systems from ever encountering them that our immune systems now don't recognize bacteria at all and can't distinguish between living or dead or harmless or pathogenic bacteria. So, the immune system stays on high alert, treating LPSs and other bacterial fragments such as dead mitochondria as if they were pathogens, triggering an attack.[29]

I like to think of our immune systems as they exist today as a poorly trained militia. In its top form, the immune system should be efficient and capable of organized, targeted attacks. Pathogens are taken care of swiftly and without much (or any) collateral damage. Sad to say, we now have a ragtag group instead, one that is anxious and overly zealous, treating everything as a dire threat. It's not their fault. They weren't taught properly.

But who was supposed to train them? Your gut buddies, of course. Your gut buddies have many ways of training the immune system. For one thing, signals between microbes influence how your immune system produces and uses antibodies.[30] This is easily observed in germ-free animals who are raised from birth without a microbiome and have profound immune system defects and a substantial reduction in antibodies.[31]

Further, you can probably picture the "whiplike" appendages on most bacteria. These are called flagellin, and they help bacteria move. We now know that flagellin also activate the immune system when facing an infection. They trigger cytokine release and instigate a signaling cascade that activates immune response genes.

When your terrain is disrupted, however, these flagellin activate immunity genes and ring the alarm to release cytokines even when you are not facing a deadly infection, leading to inflammation.[32] Remember, this alarm that is constantly going off doesn't just release cytokines in the gut. It tells the immune cells throughout the body to be on the lookout and ready to attack, which leads to widespread inflammation that runs from your gut all the way up to your brain. This neuroinflammation is the real root of the problems we are currently facing with our brains.

Even mild brain inflammation can cause brain fog and a lack of mental sharpness. Recently, it was shown that brain fog as a symptom of long COVID actually stems from neuroinflammation.[33] Side note—do you know what viruses are very good at? Causing leaky gut—the real root cause of these neurological symptoms![34]

Worse yet, long-term or more severe cases of neuroinflammation are the key drivers of neurodegeneration that occurs with age, as well as the onset and/or progression of serious neurological diseases, such as Alzheimer's disease and Parkinson's disease.[35] I'll discuss this in much greater detail later on. But for now, suffice it to say that germ-free mice do not experience any age-related increases in neuroinflammation.[36] Well, duh. That's because neuroinflammation stems from bacteria in the gut.

But the impact of neuroinflammation goes further than this. When the immune cells in your brain, which are called glial cells, receive the signal that invaders are coming, they take extraordinary measures. Glial cells are the protectors of your neurons, and they take their job very, very seriously. They nourish and support your neurons and clear your brain of waste and dead cells. They also prune away dendrites, the structures that help neurons receive information from other neurons, when they get old and weak, which allows other dendrites to strengthen themselves and take over.

Your glial cells have an intimate relationship with your gut buddies that started back before you were even born.[37] When glial cells receive the message from your gut buddies that invaders are coming, they increase their pruning process aggressively, completely cutting off connections between neurons.[38] I like to think of this as them pulling up the drawbridge to protect the castle.

This is a bold but sometimes necessary strategy in the case of a kingdom's defense, or in our case, a cell's. With the former, it leaves the castle without any access to outside supplies. In your brain, it means that your neurons are left unable to send and receive chemical messages. But these messages are the literal chemical sparks of thoughts and memories.

Remember those bacteria I mentioned earlier that were recently found in the brain? They seem to have a very specific preference for a certain type of cell called astrocytes, which work with the glial cells to support your neurons. While we don't know for sure, it certainly looks to me like they are there to help regulate the brain's immune response.[39]

We've only scratched the surface, but I hope you're beginning to see how the gut is the real root of neuroinflammation, which, in turn, is the real root of neurodegeneration and many common mental health and cognitive issues. With our disrupted inner terrains, almost all of us now have leaky gut, causing a near-constant activation of the immune system and chronic neuroinflammation.[40] It's a recipe for disaster, but it can be turned around, along with the diseases and symptoms that it causes.

CHAPTER 2

THE MANIPULATED BRAIN

As a result of our misdirected focus on sterilization rather than on fostering a healthy internal terrain, the vast majority of us have been left with dysbiosis and leaky gut. I am not exaggerating when I say that nearly 100 percent of the patients who come to see me test positive for leaky gut. By the time they get to me, it has caused dramatic and measurable changes to their brains.

These changes manifest in a wide variety of ways, from simple brain fog to neurodegenerative disease, depression, anxiety, and even addiction and a tendency to overindulge in sugary or processed foods. Yes, all of these—even eating habits—stem from the gut. When we balance my patients' inner terrain and heal their leaky guts, it is glorious to see their mental health and cognition return to their previously healthy, normal states.

If you've read some of my other books, you probably already know that the holobiome is positioned to be your most important sensory organ. This is for a few reasons. One, their bacterial genes (to say nothing of viral and fungal genes) so vastly outnumber our human genes. Two, they experience extremely rapid turnover and reproduction. And three, the holobiome is the first "organ" to make contact with the outside world, whether that's through your skin, your mouth, or your gut.

Moreover, because the holobiome has such overwhelming computing power, we (the animal) have uploaded much of our minute-to-minute decision-making as a symbiotic organism to our holobiome "cloud." This is much like how your computer's memory does not exist on its laptop. It, too, is downloaded from "the cloud."

This concept is extremely important to understand going forward, so let me give you an example. Surely you have noticed that when you go searching for certain information on your phone or you click on a particular ad, suddenly related articles, ads, and threads magically begin appearing on your phone. And the more you click on news articles with one particular point of view, the more of those views and opinions you start seeing.

The "cloud" (or algorithm) learns your preferences and gives you more of the same, and the more of these articles you see, the fewer you see with the opposite point of view. Pretty soon, you're only seeing one perspective. But you were in charge, and the computer was doing your bidding, right? Well, not exactly.

My wife, Penny, recently decided to "activate" her phone's ability to eavesdrop on her conversations. Sure enough, within minutes of talking about a place we like to visit in Italy, several posts appeared on her social media feed about all the wonderful things to do there. And, of course, when she mentioned a handbag that a friend had just purchased, ads for that particular bag magically appeared in all her feeds.

Well, of course this wasn't magic. The cloud listened, the programs installed on her phone responded, ads were generated, and Penny was given information. More important, she was presented with temptations to take action.

We now know that this same type of communication process has been going on between your microbiome and your brain for eons. That is, your thoughts and emotions and desires are really the "feeds" that are being sent to your brain from the cloud, which is your holobiome. This system was designed perfectly to create a state of homeostasis within you and your brain. But this requires a balanced terrain and an intact gut wall. When you have dysbiosis and leaky gut, the "feeds" you receive are like ads for things that are directly harmful to you but that you feel certain you want and need.

Only recently have we been able to unlock the code, or language, behind these feeds. In addition to driving neuroinflammation, your gut buddies communicate with your nervous system, including your brain, constantly. They let your brain know what's happening down in the gut and what the brain needs to do in response.

If you have a healthy, balanced terrain, your brain gets the message that everything is fine, you can relax and be happy, and it's a good idea to eat healthy, nourishing foods. This serves your gut buddies, allowing them to grow and multiply and live in a state of peace and balance, and of course, it serves you. This is an example of a healthy and balanced social media "feed."

However, when dysbiosis allows the bad bugs to overgrow, they hijack these messages and start calling the shots—telling you to feel, think, and act in ways that benefit them and not you. Bad bugs also find shockingly creative ways to tell your brain to ingest the foods and other substances that *they* want so they can grow and multiply, crowding out the friendly gut bugs even more. Spoiler alert—these are not the substances that are best for you. When these messages are hijacked, it can lead to brain fog and degeneration, as well as eating disorders, addictions, mental health issues, and more. That's a toxic social media "feed."

These messages from the gut to the brain even play a large role in determining your personality. You may be thinking, *What?* As John McEnroe would say, "You cannot be serious!" Yes, indeed, I am. So, pay attention.

As just one example, adults with a high relative abundance of bacteria called *Gammaproteobacteria* tend to be more neurotic than others, while people with a high relative abundance of butyrate-producing bacteria tend to be more conscientious! And bacterial diversity in general—the hallmark of a healthy terrain—is associated with being open and agreeable.[1] If you don't like the idea of your personality being controlled by your gut, think about it this way—the next time you find yourself arguing with your partner, you can honestly tell them, "It's not me, honey; it's my gut bugs."

Like so many of my female patients who initially come to see me for other reasons (like obesity or autoimmune disease, for instance), Kathy's medication list told a familiar story. Unrelated to her other symptoms (so she thought), she just happened to be taking two medications for depression. She had put them on her list almost nonchalantly. In fact, "depression" wasn't even on her complaint list!

When I inquired about how long Kathy had been taking these medicines, she said, "Too many years to count." When I asked if she had tried to get off them, she said once again, "Too many times to count." I told her that based on my experience with hundreds of women in similar situations, she shouldn't be surprised if, following our work together, she didn't need them anymore.

Kathy's blood work told the usual story: a low vitamin D level, low omega-3 levels, leaky gut, antibodies to wheat and gluten, and elevated markers of inflammation like hs-CRP, Il-6, and TNF-alpha. She started the Gut-Brain Paradox Program, and when I retested her three months later, her numbers had changed dramatically for the better. I saw Kathy again after six months, and when I walked into the exam room, she had a sheepish grin on her face.

"Don't get mad at me," she said, "but I weaned myself off of both of my antidepressants." I smiled. "And you won't believe how much better my brain works!" Of course I believed her. I've seen it happen so many times when you change the underlying cause of so many patients' problems: their altered terrain!

More on this later, but first let's take a look at how the gut sends these messages and exerts its control.

THE SECRET LANGUAGE OF MICROBES

The primary way your gut buddies influence your brain is through signaling molecules, which are also called postbiotics or metabolites. Don't get these confused with prebiotics, which are the foods that your gut buddies eat, or probiotics, which are the bacteria themselves! Gut buddies produce postbiotics mainly via fermentation. They are the main "language" of gut buddies, and they can influence your brain both directly and indirectly.

These postbiotics can travel from the gut to the brain via the vagus nerve, the main nerve of the parasympathetic nervous system. The vagus nerve plays an important role in your mood, immune response, digestion, and heart rate, and it is essential in maintaining homeostasis. It runs all the way from the gut to the brain, and I like to say that your gut buddies use it like a telephone wire along which they send signals upstairs about what's going on down in the gut. I'll probably have to come up with a new metaphor soon, as telephone wires are quickly becoming a thing of the past. Fiber optics? But I digress . . .

Importantly, while signals can travel in both directions up and down the vagus nerve, roughly 90 percent of them are sent up from the gut to the brain instead of the other way around.[2] Does this mean the gut is determining 90 percent of what goes on in the brain? Believe it or not, I think that's a fair estimate.

These postbiotics tell your immune system whether or not to attack.[3] And they quite literally instruct your brain on how to think, feel, and act.[4] In addition to traveling via the vagus nerve, postbiotics can also directly enter the bloodstream and/or lymph system to reach your brain or other neurons in your body.

Once they reach the end of the telephone line or a capillary, some postbiotics can cross the blood-brain barrier (BBB), a semipermeable membrane that functions similarly to the gut wall but to separate and protect the brain from the bloodstream. As an important side note, your gut buddies play a big role in helping maintain the integrity of the BBB. Just as butyrate protects the gut wall, it also helps maintain the BBB. Germ-free mice have highly permeable BBBs, and supplementing them with butyrate-producing bacteria restores the BBB to health.[5]

With leaky gut and/or dysbiosis, on the other hand, invaders can begin seeping through the BBB, all while your gut sends the message along the vagus nerve telling the brain that you are under attack. This sadly allows the BBB to become more and more permeable while the brain and the entire nervous system become more and more inflamed. In this environment, neurodegenerative diseases can thrive, along with depression, anxiety, and other mood disorders.

While there are countless postbiotics in the body, for the purposes of this book, I'll focus on those that directly impact the brain. These fall into four main categories:

Neurotransmitters

A neurotransmitter is simply a chemical that allows neurons to communicate with one another. By producing many of your body's most important neurotransmitters and/or their precursors, your microbiome controls how your neurons talk to one another—the very basis for your own thoughts. Excitatory neurotransmitters prompt neurons to share information with one another, and inhibitory neurotransmitters block these messages from being sent. They balance one another, and this balance can easily be thrown off by a disrupted inner terrain.

For instance, neurons and glial cells can only make glutamate (an important neurotransmitter) using metabolites produced by your gut buddies as precursors.[6] Cells in the intestinal tract can also produce glutamate and use it to send rapid signals to the brain.[7] Glutamate is an excitatory neurotransmitter that is responsible for sending signals between nerve cells and is involved in neuroplasticity, learning, and memory.[8] Altered levels of glutamate are associated with mood changes, psychotic disorders, and even the risk of suicide.[9,10,11]

Acetylcholine is another excitatory neurotransmitter. It is produced by bacteria, can be synthesized by neurons,[12,13] and is an important neuronal modulator.[14] Acetylcholine coordinates how groups of neurons fire together, changing how the brain responds to inputs.[15] Too little acetylcholine can lead to problems forming and recalling memories—something that far too many of us are now struggling with due to our disrupted terrain.

We often write off memory issues as a symptom of exhaustion or a mere "senior moment." Perhaps we should start referring to these as "sh*t for brains" moments, instead. Big surprise here—patients with Alzheimer's disease usually have altered levels of acetylcholine in their brains.[16]

Meanwhile, gamma-aminobutyric acid (GABA) is an inhibitory neurotransmitter that counterbalances the action of glutamate. In the brain, GABAergic neurons produce an enzyme that converts glutamate into GABA.[17] Therefore, you need healthy levels of glutamate in order to have healthy levels of GABA!

Other metabolites produced by your gut buddies, including the short-chain fatty acid (SCFA) acetate, are also a part of the GABA production process. These can cross the BBB into the hypothalamus so that GABA can be produced in the brain.[18] This is important, as low levels of GABA are linked to depression and mood disorders.[19] Having the right levels of GABA is essential for modulating neurons and neurological function.[20]

The well-known neurotransmitter dopamine, however, has effects that are both excitatory and inhibitory, and is tied to the brain's reward system. As you'll read more about later, dopamine levels play an important role in developing and recovering from addiction. Dopamine and dopamine receptors are widely distributed in the intestinal tract, and more than half of the dopamine in the body is produced by the gut.[21]

Finally, the inhibitory neurotransmitter serotonin is particularly essential, as it affects the birth of new neurons that express both dopamine and GABA.[22] Serotonin is mostly known for its connection to many common antidepressants. In fact, 90 percent of your serotonin is produced in the gut.[23] Abnormal expression and function of serotonin in the brain are both associated with depression and anxiety disorders.[24] Serotonin also acts on microglia, those immune cells in the brain, and influences levels of neuroinflammation.[25]

SCFAs

Short-chain fatty acids (SCFAs), including acetate, butyrate, and propionate, are another main category of postbiotics.[26] Your gut buddies make SCFAs by fermenting dietary (prebiotic) soluble fiber, which is usually found in plants but is also present in some animal tissues. SCFAs play countless roles in your brain. They directly affect the nervous system and enhance cholinergic neurons, which play an important role in brain function and release an essential neurotransmitter, acetylcholine.[27] The SCFA butyrate has even been shown to have antidepressant effects in animals with depression and mania.[28,29]

As you already know, SCFAs also impact the brain by acting on the immune system, modulating neuroinflammation.[30] When germ-free mice with severe microglial abnormalities were treated with SCFAs, those irregularities were reversed.[31]

Butyrate, my favorite SCFA, can also cross the BBB and enhance cholinergic neurons.[32] Butyrate also activates the vagus nerve and the hypothalamus[33] and, as mentioned, has been shown to have antidepressant effects in animals with depression and mania.[34,35]

Amino Acids

The two amino acids produced in the gut that have the biggest impact on the brain are tyrosine and tryptophan.[36] Remember the cooperative digestion between gut buddies that I mentioned earlier? The amino acid tryptophan is a great example of this. Several species of microbes produce tryptophan, which can cross the BBB and directly impact the brain and its cognition. Meanwhile, other gut buddies can ferment tryptophan to produce several different metabolites that each have their own dramatic effects on the brain.

Some of these tryptophan-derived metabolites help reduce inflammation in the central nervous system.[37] One, called indole, helps promote the birth of new neurons (neurogenesis) and strengthen the connections between neurons.[38] However, having the right balance of indole is important, as too much of it can lead to anxiety.[39] Yet another tryptophan-derived metabolite, kynurenic acid, helps regulate levels of glutamate, that neurotransmitter that impacts cognition, memory, and mood.[40]

Perhaps most important, tryptophan is also a precursor to serotonin. And tryptophan itself helps suppress inflammatory signals in the microglia.[41] It's no wonder, then, why dysregulated levels of tryptophan can lead to depression.[42] And, as you learned in my last book, *Gut Check*, glyphosate, the active ingredient in the weed killer Roundup, targets and kills the tryptophan-producing bacteria in your gut.[43] I'm getting depressed just thinking about that!

Gasotransmitters

Gasotransmitters are basically transmitters (chemical messengers) that are made of gas. There are four main gasotransmitters made in your gut—hydrogen sulfide (H_2S), nitric oxide (NO), carbon dioxide (CO_2), and hydrogen (H_2). There are others (like methane), but that's enough for now.[44]

When you experience pain, your gut buddies produce H_2S and send it to your brain to let it know that you're hurt. The H_2S then activates sensory neurons in the brain, which leads to the release of inflammatory cytokines and growth factors to heal the damage.[45] H_2S is also linked to memory. It accelerates activity in the hippocampus, the center of emotion, memory, and the autonomic nervous system in the brain. It also increases synaptic plasticity,[46] strengthening the junctions between neurons that allow them to communicate.

Meanwhile, your gut buddies produce NO through the fermentation of ammonia. NO helps regulate dopamine, which you know is tied to our reward systems in the brain.[47] It also mediates the effects of various neurotransmitters, including norepinephrine, serotonin, and glutamate. Patients with depression often have altered NO levels.[48] In addition, NO plays a role in developing the nervous system. It triggers the growth of nerve fibers, neurogenesis, and the formation of new synaptic connections.[49]

NO is also directly linked to mental health. Too much NO damages synaptic connections and the neurons themselves[50] and is present in patients with severe bipolar disorder. In fact, lithium and many antipsychotic medicines work in part by altering levels of NO in the brain![51,52]

Meanwhile, CO_2 reduces neuroinflammation by regulating cellular redox responses.[53] This is the balance between creating damaging reactive oxygen species (ROSs) and removing them with antioxidants. CO_2 also plays an important role in fetal brain development.[54] As you'll see later on, the mother's inner terrain plays a big role in the growth of her fetus's developing brain.

As I've written about before, breath holding and breath modulation techniques work in part by increasing blood and brain levels of CO_2. In fact, the reason that breathing into a paper bag helps with hyperventilation caused by a panic attack is that it causes you to breathe back in your own CO_2. Presto! Wonderous calm.

Finally, hydrogen gas is the smallest, most easily diffusible molecule in existence, and its effects as a gasotransmitter are so numerous that over 1,500 papers exist in the literature on its effects in the body. [55,56,57,58,59,60] Hydrogen is produced via fermentation by certain bacteria, but its effects on mitochondria function and brain function are paramount. Indeed, lack of hydrogen-producing bacteria has directly correlated to developing Parkinson's disease in humans.

But the importance of hydrogen goes beyond disease states. Recent reviews have highlighted hydrogen's effects on mental disorders, mood, anxiety, and depression, owing to its antioxidant and anti-inflammatory effects in the brain.[61] Patients with major depressive disorders have low levels of hydrogen-producing bacteria, as well.[62] The point of all this: Gas up!

THE GUT/MITOCHONDRIA CONNECTION

The communication system between your gut and your brain goes far beyond the "simple" act of postbiotics traveling to the brain via the vagus nerve, blood, or lymph. There is also a direct line of communication between the gut and your mitochondria, the organelles in your cells that produce energy. Your gut buddies actually exert a tremendous amount of control over your mitochondria, which have evolved from bacteria over the course of billions of years. In fact, although mitochondria live in your cells, they contain not human, but bacterial DNA.

The importance of mitochondria when it comes to the health of your brain cannot be overstated. Without functioning mitochondria, your cells, including neurons, have no way to produce energy, and so they die. As such, strong mitochondria are crucial for healthy aging, and poor mitochondrial function has been found to drive cognitive decline, neurodegenerative disorders, and neurodegeneration.[63] Changes to mitochondrial function are also linked to major depressive disorder (MDD), schizophrenia, bipolar disorder, Alzheimer's disease, and multiple sclerosis.[64]

Your gut buddies and mitochondria communicate constantly, with the messages going back and forth in both directions. People with mitochondrial disease are more prone to bacterial infections,[65] and a disruption to our internal terrain can lead to mitochondrial mutations.[66] Your gut buddies certainly have a lot to say about how your mitochondria function. For one thing, the LPSs that cross a leaky gut can create inflammation that directly harms mitochondria. In addition, your gut buddies send your mitochondria information via postbiotics that influence whether and how much energy the mitochondria can produce.[67]

Remember the serotonin that is made from tryptophan in your gut? To produce energy, your mitochondria need two cofactors made from that serotonin: nicotinamide adenine dinucleotide (NAD) and NAD phosphate (NADP).[68] In addition, SCFAs from your gut increase mitochondrial energy production, while nitric oxide (NO), which is also made in the gut, reduces it.[69]

Your gut and mitochondria also work together to birth new neurons throughout life in a process called neurogenesis. We need to continually birth new neurons throughout our lives so we can continue learning, and the reduction of adult neurogenesis is linked to memory decline, depression, and anxiety.[70]

To help birth new neurons, your gut buddies produce SCFAs to act on your mitochondria. When your gut buddies send SCFAs as messengers to the mitochondria in your neural stem cells, they tell them to replicate themselves in a process called mitogenesis and for the stem cells to differentiate into new neurons via neurogenesis.[71] Additional molecules that are created by your gut buddies, such as lactate,[72] also help promote neurogenesis. Your gut buddies also produce carbon dioxide (CO_2), which induces mitogenesis.

More neurons = better. And more mitochondria = more energy for those new neurons, so that's even more better. Perhaps this is why, in a human study of adult twins, prebiotic supplements that positively impacted the makeup of the gut biome significantly improved participants' cognition.[73] Sh*t for brains, indeed!

EXTRACELLULAR VESICLES

Another fascinating way that the gut communicates with the brain is through what's called extracellular vesicles (sometimes called exosomes). These are bilayer membranous substructures produced by various types of cells. You can think of them as little shipping containers that your cells use to send "cargo" in the form of DNA, cellular proteins, nucleic acids, and even organelles to other cells.

Sometimes extracellular vesicles contain mitochondria, which can help regulate the function of other cells, including neurons. I like to think of these extracellular vesicles as a "link" that you can click on to open up an article, a file, or a program like Zoom. There's a lot of info potentially contained in that vesicle or "link."

Extracellular vesicles can travel through your circulation system to anywhere in the body and are often exchanged between cells. Unique markers on the surface of extracellular vesicles bind to receptors on the target cells. This ensures that precious cargo isn't accidentally delivered to the wrong cell. Brilliant! Interestingly, extracellular vesicles are now being looked at as a way of potentially administering and distributing medicine in the body.[74] There are even food-derived vesicles that can deliver nutritional compounds.[75]

It turns out that in addition to our human cells, our gut buddies and even plants can also release extracellular vesicles. It's yet another part of their language, and it can have profound impacts on the body and brain. Extracellular vesicles from your gut buddies cross the gut wall and even cross your cell membranes to act directly on immune cells and neurons.[76] For example, extracellular vesicles released by our good friend *Akkermansia muciniphila* can reduce inflammation[77] and induce the secretion of serotonin.[78] Remarkably, extracellular vesicles from another gut buddy called *Lactobacillus plantarum* can have an antidepressant-like effect.[79]

Your gut buddies also use extracellular vesicles to maintain their healthy terrain—or at least, they try. To keep themselves from overpopulating, some strains of bacteria release extracellular vesicles containing toxins that kill other strains within the same genus! Meanwhile, other bacteria release extracellular vesicles containing toxins that work against competitive species.[80,81]

Notably, at least two different gut buddies (*Pseudomonas aeruginosa* and *Burkholderia thailandensis*) release extracellular vesicles with antimicrobial activities against certain antibiotic-resistant bugs, including methicillin-resistant *Staphylococcus aureus* (MRSA).[82,83] With over seventy thousand severe MRSA infections in the US each year, this is incredibly eye-opening. If only we could nurture a healthy inner terrain, those antibiotic-resistant bugs wouldn't pose so much of a threat. Our very own gut buddies would protect us from them. But then again, it's our overuse of antibiotics that has disrupted our terrain so much to begin with and made these bad bugs resistant to antibiotics at the same time. If that isn't a paradox, I don't know what is.

LOST IN TRANSLATION—THE CELL DANGER RESPONSE

Just as the right signals from your gut can tell your mitochondria and your brain that all is well, different signals—or even just a lack of signals altogether—tell them that something is amiss. This is exactly what happens when you have too few butyrate-producing bacteria because of a disrupted terrain and/or too many LPSs because of leaky gut. When mitochondria receive these abnormal signals, they shift gears from producing energy for the cell to protecting the cell from harm. They do this through what's called the cell danger response (CDR).

The CDR is a normal, healthy response to a threat, stress, or injury, consisting of three stages from the initial injury (or trigger) to healing. It can be triggered by bacteria, viruses, fungi, protozoa, or exposure to biological or chemical toxins. This is another example of a beautiful, intricate system that works well if your terrain is healthy, and otherwise . . . not so much.

Many of my fellow scientists and researchers now believe that the cell danger response is behind many of the brain conditions that are currently on the rise. The problem isn't the CDR per se, but rather the fact that our cells are never able to complete the healing cycle. For any given cell, one step in the healing cycle cannot begin until the previous step has been completed and mitochondrial function in that cell is ready for the next step. If the CDR is triggered over and over by LPSs, toxins, and overgrowing pathogens, the cycle, unfortunately, never completes.

Let's take a closer look at how it works.

Stage 1

Stage 1 of the CDR is the activation of the immune system—inflammation—to instigate healing. When cells get the message that something is wrong, the first thing they do is start leaking ATP—that's right, the very energy source that powers your cells. Inside the cell, ATP provides energy, but extracellular ATP (eATP), meaning ATP that's outside of the cell, is a signaling molecule to other cells that warns them of danger.[84] The damaged cell literally releases ATP to signal (or yell) to neighboring cells, "Hey, I just got bit by a shark! Don't try to save me. Get out of the water, save yourselves!"

As the cell leaks ATP, your mitochondria go through dramatic changes. In fact, they change from typical healthy mitochondria called M2 to pro-inflammatory mitochondria called M1.[85] Interestingly, the same change happens in your brain's glial cells when they are treated with pro-inflammatory triggers such as LPSs.[86]

At the same time, eATP triggers the cells to harden their membranes with patches of cholesterol and ceramides to protect themselves.[87] This cuts off communication with surrounding cells. It is a necessary step, but it leads to a temporary—or what should be temporary—decrease in organ function.

Bear with me for a quick detour here to talk about ceramides.

These lipid molecules have become a popular skin-care ingredient because they play an important role in the skin's barrier function.[88] They may be good for your skin, but they're not so good for your mitochondria. In fact, ceramides negatively impact mitochondrial function, primarily by suppressing levels of orexin, a neuropeptide produced in the neurons that promotes wakefulness, during the day, and melatonin, a hormone that helps regulate circadian rhythm, at night.

We need orexin and melatonin to help modulate inflammation and protect mitochondria from oxidative stress, which is an imbalance between harmful free radicals and beneficial antioxidants.[89] No wonder an increase in ceramides is implicated in a number of psychiatric conditions, particularly bipolar disorder and schizophrenia.[90]

By the way, do you know where the majority of the melatonin in your body is produced? In your gut, of course. And melatonin also helps protect the gut lining . . . if you have enough of it.[91] Just another example of how disrupted terrain and leaky gut are a one-two punch to your brain. If your cells are constantly going through the cell danger response cycle, you'll have more inflammation in the mitochondria and less melatonin to protect your gut wall. Oh, and your circadian rhythm will be thrown off, too. In older adults, there is a positive association between melatonin-containing food consumption and life satisfaction, psycho-emotional state, and cognitive function.[92]

Luckily, butyrate can inhibit the production of ceramides. But wait, this cycle may not have started to begin with if you had enough butyrate-producing bacteria to tell your mitochondria that all was well. So, scratch that.

Stage 2

Once the damage from the initial stress or injury is healed during stage 1, the cell danger response cycle moves to stage 2. At this point, communication is reestablished between the cells that survived stage

1 unscathed. But many cells don't make it this far. The hardening of their cell membranes and damage to their mitochondria lead to their death. During stage 2, stem cells are recruited to replace these lost cells.

Stage 3

During this stage, the stem cells that have been recruited to replace dead cells differentiate into the type of cells that are needed and establish communication with neighboring cells. At the same time, their mitochondria return to their healthy form as M2 mitochondria. Finally, the cells stop leaking ATP because they are no longer under threat. When eATP levels decrease, the healing cycle is over.

CHRONIC ILLNESS OR LACK OF HEALING?

In a perfect world, the CDR would only be triggered in the case of a rare dangerous bacterial infection or trauma. The cycle would be completed, healing would take place, and all would return to normal. But we live in a far-from-perfect world.

With our disrupted terrain and poorly trained immune systems, our body's alarm systems are constantly going off. Cells can rarely make it through the CDR before the alarm goes off yet again. Healing never takes place, and our cells get stuck in a loop of inflammation and mitochondrial disfunction.

In addition, eATP is released in your gut during times of rapid bacterial population growth. In other words, when your terrain is disrupted and certain species overgrow, your gut sends the danger signal via eATP. This triggers the CDR to begin.[93] Oh my!

We are now seeing that in the case of most chronic illnesses, mitochondria are stuck in a stage of the CDR that was meant to be temporary. The CDR does not turn off until eATP levels return to normal. If this doesn't happen, the cell remains focused on survival rather than regular functioning.

Over time, sustained changes in mitochondrial function can lead to structural changes in tissues and organs that can make full recovery more difficult.[94] When this persists, it leads to the impairment of organ systems, negative changes to the microbiome, and chronic disease.[95,96] Because of its impact on the brain, this can change human thought and behavior.[97]

In fact, many chronic illnesses can be classified based on which stage of the CDR is blocked. For instance, blocks in completing stage 1 can cause chronic inflammatory disorders, blocks in stage 2 can lead to proliferative disorders such as atherosclerosis and cancer, and blocks in stage 3 can lead to neurodevelopmental, affective, neuropsychiatric, and neurodegenerative disorders.

In particular, there is strong evidence of the link between the CDR and autism. In one study of ten boys between the ages of five and fourteen, half of them received a single dose of antipurinergic therapy, which reduces levels of eATP. The other half received a placebo. All of the participants who received the antipurinergic therapy experienced improvements in language, social interaction, restricted interests, and repetitive movements. Two children who had been nonverbal spoke their first sentences! No improvements were seen in the placebo group.[98]

I believe this offers a lot of hope for those suffering from chronic illness. Even without these therapies, the Gut-Brain Paradox Program has helped many patients recover from long-term disease. One way I help them do this is by dramatically increasing their intake of polyphenols, important compounds in our food. When mitochondria are stuck in the CDR and become dysfunctional, they start producing reactive oxygen species (ROSs), increasing inflammation. Polyphenols scavenge for ROSs. This is just one way we will reduce inflammation and interrupt the blockage of the CDR to promote healing.[99]

* * *

I hope you are beginning to appreciate the delicate communication systems between our guts and our brains and see that many of the brain conditions we are suffering with are the unfortunate results of a profound disconnect. The good news is that we can take back control of these systems and, as a result, take back the health of our brains.

CHAPTER 3

YOUR HUNGER HAS BEEN HIJACKED

With your gut buddies sending so many different types of messages to your brain, is it really a stretch to believe that they are the ones in control of how you think and feel? What about your behaviors and habits? I'm firmly convinced that they are running all of it.

Your gut buddies are the proverbial puppet masters pulling the strings, and you are merely the puppet. Sorry. By the end of this book, I hope to recruit you to this way of thinking. Or I should say that I hope to recruit your gut buddies to this way of thinking, since they're the ones who are really in charge!

Before we get to other behaviors and illnesses, let's start by taking a look at how your gut buddies impact your hunger and eating habits.

YOUR MICROBIOME AND YOUR METABOLISM

We have known for a long time that the makeup of your microbiome plays an enormous role in dictating your weight. We have seen evidence of this in many studies featuring my favorite gut buddy, *Akkermansia muciniphila*. The more of this gut buddy you have, the leaner you are likely to be—and vice versa.[1] In fact, even with the new injectable

GLP-1 agonists, your weight loss response to the drug depends on the types of bugs in your gut.[2]

Perhaps most compelling are studies on people who have undergone fecal microbiota transplants and go on to gain or lose weight regardless of dietary changes. This happened to one woman who was suffering from a treatment-resistant *Clostridium difficile* infection and received a fecal microbiota transplant from an overweight donor. Before the transplant, the patient's weight was stable at 136 pounds. Sixteen months after the transplant, she had gained 34 pounds, putting her at 170, despite the fact that she had made no dietary changes and had normal thyroid and cortisol panels. Even after going on a medically supervised liquid diet and exercise program, she continued to gain weight. Three years after the transplant, this patient weighed 177 pounds.[3]

While cases like this are bizarre, they are not all that surprising, and we are now beginning to understand exactly how messages from your gut buddies influence your weight. As this transplant patient clearly shows, weight gain (and loss) is not just about what you eat, or even how much. If that doesn't convince you, think about this: Young adult mice have 40 percent more total body fat than their germ-free counterparts, even when they are fed the exact same diet.[4]

Clearly, your microbiome is driving your metabolism. But why am I talking about weight in a book about the brain? Excellent question. It turns out that metabolic disturbances stemming from the gut lead to cognitive decline. It's all about that dreaded neuroinflammation.

When you have leaky gut and dysbiosis, widespread inflammation reaches the brain. This neuroinflammation changes the ways you act, think, and feel and drives neurodegeneration (something I'll discuss in much greater detail later on). As you're about to see, metabolic dysfunction is often something of a middle entity (I was tempted to say middleman, but I trust that it's gender neutral), representing an intermediary step between dysbiosis and leaky gut and the resulting changes in the brain.

Further, it turns out that you are not as in control of your food-related decisions as you may think. Your gut buddies send messages to your brain telling you to eat the foods *they* want you to eat—the ones they can use to be fruitful and multiply. Disturbing? Yes. Liberating? I hope so. You are not a bad person if you can't seem to break your junk food habit or lose weight! You might just have a "bad" microbiome. Further, even if you stay at the same weight, a healthy microbiome can do wonders for your mental and overall health.

ENDOTOXEMIA AND METABOLIC DISEASE

Let's start by looking at the striking relationship between your gut biome and adipose or fatty tissue. Now, adipose tissue is not just stored fat. It is an organ with functions related to your neurological, immune, and endocrine systems. So, when you have problems with your adipose tissue, the results are greater than just the accumulation of additional fat. When the cells in your adipose tissue increase in size or over-proliferate, it leads to chronic inflammation, which causes endothelial cell dysfunction. This makes the BBB leaky, leading to cognitive decline.[5]

But what causes this type of issue with adipose fat? Another good question. Earlier, I mentioned the many different metabolites that your gut buddies produce out of the amino acid tryptophan. Several of them act directly on adipose tissue to regulate energy expenditure and insulin sensitivity.[6]

But the primary way that the gut biome leads to fat, inflammation, and metabolic dysfunction is through LPSs—often stemming from too much saturated fat. In general, the more saturated fat we eat, the higher our concentration of plasma LPSs.[7] Endotoxemia is the condition that occurs when you have high levels of LPSs in the blood, leading to widespread inflammation and metabolic syndrome. Infusions of LPSs, triggering endotoxemia, lead to weight gain and insulin resistance.[8] Spoiler alert—endotoxemia is also a risk factor for cognitive impairment.[9]

As one study shows, dysbiosis stemming from a diet high in saturated fat is the root of metabolic disturbance that leads to cognitive decline.[10] Hold the phone. This means that your diet in and of itself is not making you fat. Well, we already knew that. But in this case, a diet high in saturated fat is causing dysbiosis, and it's that disrupted terrain that's making you fat.[11] This is a subtle but incredibly important distinction. And, yes, this means that we can—and will!—fix the problem in much the same way. By healing the gut, we can reverse the metabolic conditions that are impacting your brain.

However, the makeup of your gut has a lot to say about whether or not and to what extent a high-fat diet leads to endotoxemia to begin with. In one study, a group of mice with saturated fat–induced endotoxemia had their gut bugs wiped out with antibiotics and then transplanted with gut buddies from a healthy group of mice. They were still fed the high-saturated-fat diet, yet they experienced reduced levels of endotoxemia, inflammation in their fatty tissue, and insulin resistance. This shows that the makeup of your gut biome has a direct influence over your fatty tissue and metabolic health, regardless of what you eat.[12]

But don't forget why a high-saturated-fat diet leads to an increase in LPSs in the first place. Because it changes the makeup of your microbiome. Endotoxemia is always the result of disrupted terrain leading to leaky gut. After all, your gut wall needs to be leaky in order for those LPSs to sneak through. Want proof of this? People with obesity have bacteria in their adipose fat, while people without obesity do not. Where do these bacteria come from? A leaky gut.[13]

So, it makes sense that a healthy terrain would be able to sustain a high-saturated-fat diet, especially during a temporary study, without causing endotoxemia. However, over time, a high-saturated-fat diet can disrupt that terrain. There are specific gut bugs, such as *Bacteroides fragilis*, that love saturated fat. These are bad bugs, and when you have a lot of them, they produce tons of LPSs. When you eat a lot of saturated fat, these bugs multiply, and your LPS levels go up exponentially. Hello, inflammation, weight gain, and cognitive impairment!

Even worse, when you have a lot of these bugs, they send messages to your brain telling you to eat more saturated fat so they can continue growing. (More on this in a moment.) Thus begins a vicious cycle of increased saturated fat intake causing more and more dysbiosis, causing more and more leaky gut and LPSs, causing more and more weight gain. At the same time, the increased abundance of *B. fragilis* send more and more signals to your brain, telling you eat more saturated fat!

Again, your dietary choices are not fully under your control— especially if you have a disrupted terrain. Once you start to shift your microbiome to a more balanced state with the Gut-Brain Paradox Program, your brain will start receiving different signals, telling you to eat healthy foods that allow your good gut buddies to grow and thrive! This will make all the difference. As one recent study summarized, diet-induced changes in endotoxemia can bridge the gap between food intake and metabolic disease.[14] Once again, changing the microbiome will allow you to change your metabolic function.

For example, when one type of endotoxin-producing bacteria, *Enterobacter cloacae*, was isolated from a morbidly obese person's gut and transplanted into germ-free mice, the mice then developed obesity and insulin resistance. Simultaneously, the human donor spent twenty-three weeks on a diet of rice gruel (congee), traditional Chinese medicinal foods, and prebiotics. The endotoxin-producing bacteria in their gut decreased in relative abundance from 35 percent to an undetectable amount. This led to dramatic weight loss, reduced inflammation, and recovery from hyperglycemia and hypertension.[15] Not bad!

One small change that can make a big difference here is the use of alpha-linolenic acid (ALA), an essential fatty acid found in perilla oil and flaxseed oil that significantly tamps down the inflammation and endotoxemia that's associated with a high-fat diet.[16] How does it do this? Through the microbiome, of course. ALA keeps bad bugs from overgrowing and feeds the good bugs that are anti-inflammatory and help protect your gut wall. This has been found to be just as effective as metformin, a common diabetes drug, in treating insulin resistance![17] ALA is just one of many beneficial compounds that we now know are so good for us simply because they are good for our guts!

Here's another one. We've known for a long time that flavonoids, a type of polyphenol, are beneficial. For instance, a flavonoid called xanthohumol (XN) and its semisynthetic derivative tetrahydroxan-thohumol (TXN) improve high-fat-diet-induced obesity and meta-bolic syndrome in mice by reducing levels of inflammation-inducing microbes.[18] Once again, the positive changes here are all coming from the gut.

By the way, guess where that XN that improved the gut biome came from? Hops. That's right, the stuff in beer. Hops have been used in beer since the ninth century. Have you ever wondered why? Until now, I hadn't really, either. But this is just one example of how our wise ancestors somehow knew how to take care of their gut buddies so their gut buddies would take care of them. We have lost this innate wisdom, and our brains are paying the price. More on this, sadly, to come.

YOUR MICROBIOME IS MAKING YOU HUNGRY

I have been saying for years that you are not what you eat—you are what your gut buddies digest. But even I didn't realize how deeply this connects to your own feelings of hunger—until now.

Your gut buddies need to digest or ferment specific foods in order for them to grow and multiply. But, of course, they only have access to the foods you eat. To survive, over many millions of years, they have evolved to make sure that you eat the foods *they* want—the ones that allow them to thrive. If you're lucky, you have the right makeup of gut bugs, and the foods that help them thrive are also good for you. Together, we'll make sure that is the case going forward.

While it has long been accepted that feelings of hunger come from the feeding center of the human brain, it turns out that it's actually your gut buddies that are making you feel hungry. In other words, when you feel compelled to eat, it's not because you the human are hungry. It's because your gut buddies are. This is one reason that people who eat only prebiotic fiber—which humans cannot digest but can be digested by our gut buddies—do not feel hungry. It's because their gut buddies are satisfied and are not sending a hunger signal to the brain.[19]

Your gut buddies have multiple ways of sending hunger signals to the brain. Remember earlier when I said that if you don't eat enough of the foods your gut buddies need to proliferate, they will resort to eating the gut lining? It turns out that this is not only their way of getting something to eat, but also one of the ways they signal to your brain that you are hungry![20]

In an exceptionally well-designed system, there are also gut buddies that exist specifically in order to detect which nutrients are missing in the body. Those bacteria then produce metabolites that travel to the brain and tell you to eat more of the foods containing those nutrients. Is this system the root of food cravings? It's tempting to say yes, and that if this is the case, we should always give in to those cravings. After all, a craving must mean that our gut buddies are trying to tell us that we're deficient in an important nutrient. Right?

This would certainly be the case if we had a perfectly healthy terrain. This system was designed so that you would crave the exact foods your body needed. Brilliant! But what if the terrain is disrupted as, sadly, so many of ours are? Then this craving system gets hijacked, and bad bugs signal for foods *they* want, which unfortunately aren't always the best for you.

For example, simply increasing the abundance of two bacteria that send these craving-inducing metabolites can suppress protein cravings and increase the desire for sugar.[21] So, just imagine what happens when the balance of your inner terrain is completely thrown off! I think it's safe to say that right now, you can't trust your cravings. So much for "intuitive eating."

It's more than likely that your hunger signals have been hijacked. But once we heal your gut, I promise that you will start craving foods that are better for you and for your gut. No more fighting food cravings. It will actually benefit you to give in to them instead!

For now, however, instead of listening to cravings, it's much more important to eat foods that we know will nourish your most beneficial gut buddies. This will help restore your hunger signals. Primarily, this means lots of prebiotic fiber.[22] When people eat a type of prebiotic that humans cannot digest but that gut buddies love to ferment, it positively reshapes the gut biome, leading to weight loss and improvements in metabolic and cognitive function.[23]

When you don't eat enough of these gut-buddy-friendly foods, certain species of bacteria die off, allowing others to overgrow. For example, a diet low in dietary fiber leads to a decrease in the gut buddies that ferment plant polysaccharides. Even worse, diets high in saturated fats and simple sugars lead to a reduction of gut buddies that produce SCFAs and a rise of pathogenic bacteria like *Escherichia coli*.[24] Again, this creates a vicious cycle, as those overgrowing bad bugs can hijack your hunger signals and manipulate you into eating more of the foods they need. This, of course, allows them to grow and crowd out the good gut buddies even more!

We evolved alongside our gut buddies, so just as our diets help shape our inner terrain, our inner terrain also exists to meet our own nutritional needs. This is why different species of animals with different dietary needs have radically different gut biomes.[25] For example, tigers, lions, and wolves have predominantly meat-eating bacteria, while horses, cows, and sheep have far more plant-eating bacteria. As we are omnivores, humans have both meat- and vegetable-eating bacteria—at least we should.[26]

THE WESTERN DIET HAS KILLED OFF THE INNER TERRAIN

The Western diet is so unfriendly to our gut buddies that it's no exaggeration to say that we have completely destroyed our inner terrain. When eight healthy people were put on a typical Western diet for one month, their endotoxin levels went up by 71 percent.[27] In one month! Imagine what a lifetime of processed foods is doing to your gut and therefore your brain.

I'll tell you this—one thing it is doing is making you hungry all the time. Chronic exposure to LPSs from a Western diet can induce leptin resistance. This is when the brain cannot respond to leptin, a hormone that suppresses appetite and increases energy expenditure. So, you eat and eat and your brain never gets the signal that you are full.[28]

This is another one-two punch, as you're eating a Western diet to begin with, which—in addition to other problems we'll address—doesn't have enough of the dietary fiber that your gut buddies need. We now rely on processed and even ultraprocessed foods that have been broken down into individual components and are essentially predigested. And in that predigestion process, we have discarded the prebiotic fiber your gut buddies need. We are bypassing our gut buddies' role in digestion completely and giving them nothing to eat! This is the exact opposite of how you should nourish a healthy terrain. As a result, your gut buddies keep sending the signal to the brain telling you to eat more, all while your brain cannot respond to the leptin that's trying to tell it to feel full. What a mess.

In addition, the Western diet is characterized by an excessive intake of fructose, which disrupts our inner terrain, increases intestinal permeability, and causes endotoxemia.[29,30] Big shock—human studies show that fructose consumption is associated with a higher risk of Alzheimer's disease and dementia.[31] And you already know about saturated fat (a hallmark of the Western diet) and its connection to endotoxemia, fatty tissue, and cognitive dysfunction. Following this diet has shifted our gut biomes away from that beautiful terrain that was meant to nourish omnivores and toward something that is akin to Frankenstein's monster.

This "Western diet microbiome" is dominated by bacteria that ferment simple sugars.[32] And it is these bacteria calling the shots, telling you what and how much to eat. Again, if you are overweight or suffering from diabetes or prediabetes, it's not your fault. Your signals are simply crossed. But we are going to fix that.

WHAT BUILT FRANKENSTEIN'S MONSTER?

On top of following a Western diet, we are now exposed to a host of environmental toxins that destroy our inner terrain and directly impact our brains. I've written about many of these before, but as we keep learning more and more about how these toxins are hurting us and our guts, I would be remiss not to include a mention of them here, specifically regarding how they impact the brain.

Glyphosate

To start, the herbicide glyphosate kills off many of our most beneficial bacteria, including those that produce SCFAs and tryptophan. Meanwhile, the bacteria in our guts that are glyphosate-resistant just happen to be pro-inflammatory.[33] There are specific bacteria that are especially susceptible to glyphosate, such as *Lactobacillus*, which breaks tryptophan down into indole, which, as you read earlier, promotes neurogenesis and strengthens the connections between neurons.[34] Besides being the precursor for your feel-good hormones, tryptophan also activates gut wall repair and signals to your white blood cells to chill out.[35] You don't want to grow old without this!

Glyphosate also kills off the bacteria *Ruminococcaceae*. Reductions in this gut buddy are associated with Parkinson's disease,[36] schizophrenia,[37] and depression.[38] *Ruminococcaceae* produces several important metabolites, including L-glutamate. In addition to other important functions, L-glutamate is a precursor to GABA. Reductions in GABA are associated with depression,[39] anxiety,[40] and premenstrual dysphoric disorder.[41] Don't worry, there are other bacteria that can produce GABA, such as *Bacteroides* and *Lactobacillus*. Oh, wait. Both of those are highly susceptible to glyphosate, too.[42] Never mind.

Antibiotics

Then there are antibiotics, which are ubiquitous not just in doctors' offices, but also in our livestock and even our crops.[43] Twenty-five percent of antibiotic prescriptions are considered inappropriate or unnecessary, and the average American is prescribed a course of antibiotics every six months.[44]

These drugs represent a huge threat to our inner terrain and to our brains. Just one course of antibiotics can cause huge changes to the terrain that last at least two months and up to two years.[45] Even then, patients' microbiomes look profoundly different than they did before taking antibiotics, with a lot of gut buddies—and the beneficial compounds they produce—completely missing.[46,47] Each time a patient is exposed to the same antibiotic, that disruption of the terrain gets more severe, and it becomes increasingly difficult to return to equilibrium.[48,49]

These changes to our terrain have a direct impact on our brains. For one thing, antibiotics can inhibit mitochondrial function.[50] Remember, mitochondria are actually ancient engulfed bacteria.

Certain antibiotics also damage and even kill neurons, creating behavioral and neurological issues such as depression and anxiety.[51] In addition, recurrent use of antibiotics in childhood is associated with an increased risk of cognitive impairment in middle and old age,[52] and antibiotic use in middle-aged women is associated with cognitive decline seven years later.[53]

Proton Pump Inhibitors

But antibiotics aren't the only drugs that damage our guts and our brains. Proton pump inhibitors (PPIs) and other acid-blocking drugs work by reducing levels of stomach acid. The problem is that acid is one of our defense systems against pathogenic bacteria. Without that acid, we are more likely to suffer from leaky gut and inflammation.

In addition, proton pumps don't merely exist to help produce stomach acid. They are a part of the cell that moves protons across a membrane. Well, moving protons across a membrane is part of how your mitochondria make energy. Yes, your mitochondria have their own proton pumps. And PPIs inhibit these, too, slowing down your cells' energy production. No wonder, then, that PPIs are associated with brain fog, cognitive slowdown, and dementia.[54]

Sleep Aids

Prescription sleep aids can also be harmful to the brain. These drugs work by artificially stimulating production of GABA, which, you recall, needs ingredients, or precursors, made by your gut buddies. Artificially triggering GABA production taxes your gut buddies and throws off your inner terrain.

As such, prescription sleep aids are connected to an increased risk of dementia. Men over sixty-five who use sleep medications have a 3.6 times increased risk of developing Alzheimer's disease compared to those who do not use sleep medications.[55] And, not to scare you any further, but general anesthesia for surgery does the same thing.[56]

Endocrine Disrupters

Finally, endocrine disrupters that mess with our hormone systems, like bisphenol A (BPA), create direct changes to the microbiome, including killing off bacteria that produce SCFAs.[57] You already know what this does to your gut and therefore your brain.

* * *

I hope you are beginning to see how much control your gut buddies have over not only your cognitive function, but also your choices and behaviors. After all, if they can get you to eat the foods they want, what else can they get you to do? The answer is: plenty. We'll explore this more later on.

CHAPTER 4

THE DOSE MAKES THE POISON

When I was a teenager, I was horribly allergic to ragweed. After I had suffered through several miserable fall seasons, when ragweed pollen is at its highest levels, my parents took me to get allergy shots. As you may know, allergy symptoms are just the side effects of the body attacking an allergen (in my case, ragweed pollen) that it views as an unknown invader. In an allergic patient, preformed IgE antibodies bind to the allergen and prompt the release of histamines from a type of immune cell called a mast cell. Histamines prompt the runny nose and watery eyes of allergy sufferers at best and can lead to death at worst.

For instance, peanut allergies involve the IgE antibody to the peanut lectin.[1] Huge numbers of people carry that antibody. Yet, when I was growing up in the 1950s and '60s, seemingly no one had a peanut allergy. We brought peanut butter and jelly sandwiches (on Wonder Bread) to school every day, they passed out free peanuts on airplanes, and we devoured them at the ballpark. Now, if a kid brings their peanut butter sandwich to school, half of the class has to whip out their EpiPen. So, what happened?

If you've gotten this far, you already know the answer. Back then, most of us had a great inner terrain and a solid gut wall. Broad-spectrum antibiotics didn't exist yet, nor did glyphosate or BPA. You get the point. Your immune system was educated by your gut buddies and knew that that peanut lectin wasn't a big deal, so it didn't overreact.

Now, the bad bugs rule your gut, your gut wall is breached daily, and your immune system is constantly on high alert. No wonder we have a problem with allergies like never before.

But back to my allergy shots. These work by essentially training the immune system to recognize the allergen using tiny amounts injected into the patient at a time so that it no longer feels threatened and therefore does not attack. The immune system learns to tolerate the presence of the antigen little by little without pulling the trigger.

The key to allergy shots is that you're not overwhelming the body with a huge amount of the allergen. When you inject a tiny bit over and over, the body becomes more and more tolerant to that antigen. Eventually, the immune system gets the message and stops freaking out every time it comes across that allergen. And, voila, your allergy symptoms are gone.

What does this have to do with your gut, and therefore your brain? Besides the connection between the gut and the immune system, if you had asked me this question before I started doing the research for this book, I would have probably said: not a whole lot. But then I uncovered some research that made me rethink everything I used to believe about one of the most toxic gut-and-brain-destroying things out there—endotoxins, aka LPSs.

For years, I've been writing about how LPSs are causing leaky gut, dysbiosis, autoimmune conditions, and on and on. And they are. But now I realize that this is only because in our post-Pasteur, sterilization-obsessed society, our immune systems have not been exposed to LPSs (or much bacteria at all) and are essentially treating them—and everything that looks like them—as allergens. We need to be exposed to the right amounts of bacteria and LPSs so our bodies learn to recognize and no longer be afraid of them. Just like with my allergy shots, and as a Swiss physician named Paracelsus said way back in the 1500s—it's the dose that makes the poison.

THE BIG BRAIN THEORY

Bear with me—I promise to get back to LPSs in a moment—but first I want to spend a little bit of time talking about how we humans got our big, impressive brains to begin with. It really came down to us being able to consume and utilize a lot of glucose, and we did this by harnessing fire. See, no animal, including humans, has the ability to digest the cell wall of a plant. We need bacteria to do it for us via fermentation.

Even termites themselves can't digest wood. They have their own microbiome to ferment it for them. And, gee, I wonder why termites want to eat that wood, then. Because their gut buddies are calling the shots!

In general, mammals have three possible places in their bodies where fermentation can occur: the foregut, midgut, or hindgut. And most animals have complex digestive systems to break down these plant walls. Foregut fermenters like cows and sheep have multiple stomachs. These are fermentation vats. They semi-ferment the food in one stomach, regurgitate it, chew it up again, and swallow it back down to be fermented further. Charming, I know.

Meanwhile, great apes are midgut fermenters, meaning they ferment most of their food in their small bowels. Have you ever wondered why muscular apes have such big bellies? That's their fermentation vat, showing where their food is being digested.

We humans have evolved to be hindgut fermenters. We ferment most of our food in the colon, and our digestive systems are much simpler than those of a great ape. They're surprisingly similar to the digestive system of a dog. In fact, dogs' microbiomes are also quite similar to humans'![2] But we evolved from apes, not dogs, so why did our digestive systems become so simple?

The answer is fire. When humans harnessed fire and began using it to cook our foods, we no longer needed such a vast and complex fermentation system to break down cell walls. Cooking our food did that for us! This not only allowed us to evolve with a simpler digestive system, but also to access a completely novel fuel source that great apes could not digest: tubers.

Tubers are the roots of a plant. They're actually where plants store their energy as starch. Starch is simply glucose molecules strung together in a long chain. In fact, storing their valuables underground in this way helps plants survive during the winter and keeps them from being eaten. Does this sound familiar? As soon as humans were able to cook and therefore digest tubers, we began storing them to cook and eat during the winter, too!

When humans began cooking tubers, we became the first animals to break down plant material without the help of bacteria. We didn't need such a complex fermentation system in our guts anymore because we could directly absorb the glucose from cooked tubers and use it as a fuel source. This led to two major evolutionary changes— the simplification of our digestive systems and the growth of our big brains.

Contrary to what many people think, our brains didn't get big while our guts shrank because we started eating meat. If that was the case, then big cats like lions, tigers, and cougars (oh my!) ought to have monstrously big brains. Spoiler alert: Their brains are small. But our big brains now use 80 percent of our bodies' glucose.[3] This is made possible by cooked starches.

Ever since the invention of fire, tubers became a universal mainstay of the human diet. Not only were they rich in glucose and nutrients, but they were some of the only vegetables that could survive without refrigeration. Back in the day, everyone had a "root cellar" in which to store their tubers in the winter.

As a side note, this is one reason I believe that eating only raw food is a very bad idea for your brain. It is simply impossible to extract enough glucose from raw food with our current digestive system to support our big human brains. Many raw foodists try to compensate and consume enough calories by relying on raw fruit. But, as I mentioned earlier, the high levels of fructose in fruit cause leaky gut and dysbiosis.

So, what does all of this have to do with LPSs? Well, as you know, tubers grow underground. That dirt or soil they grew in was historically absolutely swimming in bacteria. When we dug up those tubers, much of those bacteria clinging to them was alive, and some of them were dead. Those tubers were also covered in plenty of LPSs.

Back to my own childhood for a moment. We had a garden out back, and as a kid, I often went outside, yanked a carrot out of the ground, brushed off the dirt, and ate it. Did that carrot have a little residual dirt on it, which included some LPSs? You bet. Did that hurt me? Not one bit. And I'm now learning that it actually helped more than I ever realized.

As humans throughout history ate tubers and consumed small amounts of LPSs along with them, we were consistently exposed to low doses of LPSs. They were like allergy shots, teaching our immune systems not to attack. But as we've completely changed the way we eat over the past fifty to one hundred years, we are no longer getting our "allergy shots." Our immune systems no longer recognize LPSs or the bacteria they came from, and our big brains are now paying the price.

THE HYGIENE HYPOTHESIS

Back in the 1980s, when the number of childhood allergies was dramatically increasing, a doctor named David Strachan put forward a theory that he called the hygiene hypothesis. The idea was that early childhood exposure to a variety of microbes protected us from allergens in a method that's sort of like nature's own allergy shots. The bacteria we were exposed to trained our immune systems to recognize and tolerate other microbes, as well.

By the 1980s, when this hypothesis started gaining traction, the world had already become far more sterile than it had ever been. This trend toward sterilization may have started with Pasteur, but it really picked up steam at the turn of the twentieth century as the industrial revolution brought about improvements in sanitation and clean water. Of course, this helped reduce childhood deaths from pathogens like cholera and typhoid. But Strachan rightly noted that all this cleanliness was not helping to decrease the incidents of childhood allergies and asthma. If anything, it was the opposite.

Now, of course, we go to even greater lengths to sanitize our environments. In addition to antibiotics, hand sanitizers, and cleaning products, our current farming practices kill off the bacteria we would naturally be exposed to, including those that should be in the soil our food is grown in. Our homes are sterile, our hands are sterile, our soil is sterile, and therefore our food is sterile. Never mind the fact that the bacteria in soil used to help plants uptake nutrients from the soil. With dead soil, our food is less nutrient dense.

In addition, the result of all this sterilization is that our immune system no longer receives an education about bacteria as a whole. So, it views all bacteria that crosses the gut wall as a threat—whether it's alive, dead, or, in the case of LPSs, in fragments. Imagine my surprise when I realized that small doses of LPSs can be very beneficial as a way of inoculating the immune system, just like allergy shots!

This concept has actually been around for a while. A study from 2002 showed that environmental LPS exposure was inversely associated with the incidence of childhood asthma.[4] This, of course, reflects the data behind the hygiene hypothesis.

But what about the brain? We know that high doses of LPSs transform glial cells in the brain to become pro-inflammatory. This can lead to neurodegeneration.[5] However, recent studies have shown that training the immune system with low doses of LPSs can lead to positive changes in the brain that help prevent damage to neurons. In fact, preconditioning glial cells with low doses of LPSs transforms microglia to become anti-inflammatory[6]—the exact opposite of what happens with higher doses!

Lo and behold, in the right amounts, LPSs can be neuroprotective. Continuous low-dose oral administration of LPSs can help prevent dementia[7] and suppress the cognitive decline associated with Alzheimer's disease.[8] It has even been suggested that oral administration of LPSs be used as a novel treatment for dementia.[9]

LPSs can also reduce neuroinflammation by limiting the release of cytokines. They do this by enhancing the response of T cells, a type of immune cell.[10] Amazingly, LPSs can also activate SIRT1,[11] an enzyme in the cell nucleus that repairs and protects DNA from damage. SIRT1 is deeply involved in brain function and aging.[12]

Perhaps most shocking, when an animal is LPS tolerant, meaning they are consistently exposed to low doses of LPSs, they are less vulnerable to infection.[13] Consistent exposure to low-dose LPSs strengthens the immune system overall.

But remember, our uneducated immune systems don't only attack LPSs, whether or not they are an actual threat. That would certainly be bad enough. They attack all bacteria that exist in locations outside of the colon, where they should be.

For example, when unnaturally high doses of LPSs were injected into the colons of mice, it did not lead to inflammation. This is because bacteria and therefore LPSs are expected and familiar in the colon. There is no reason for the immune system to attack. But when high doses of LPSs are injected into the small intestine, instead, they do indeed cause inflammation along with tissue damage.[14] This is because LPSs aren't supposed to be there. They are unfamiliar and are therefore seen as invaders and attacked.

You may be thinking—isn't it a good thing to attack bacteria that's outside of the colon? Yes and no. One problem is that, as I mentioned earlier, our mitochondria contain bacterial DNA. Anytime our immune system encounters extracellular mitochondria, they attack. This drives neuroinflammation.

But wait, why would mitochondria be outside of the cell? Another good question. When the mitochondria inside of your cells become damaged to the point of no return, the cell dies and explodes in a process called apoptosis. This process throws cellular debris, which includes mitochondrial walls, into your bloodstream. Because of their bacterial DNA, those mitochondrial cell walls might as well be LPSs or pathogenic bacteria as far as your untrained immune system is concerned. So, it attacks, leading to neuroinflammation.[15]

While all cells must eventually die, not all cells die through apoptosis. There is another, far preferable way for cells to die. It is called autophagy, which literally means to "self-eat." In this process, all pieces of the cell, including the mitochondria, are recycled into new cells. In this case, there is no extracellular mitochondria, and nothing for the immune system to react to. Keeping your cells and their mitochondria healthy by restoring your inner terrain will help cut down on apoptosis, reducing inflammation. We will do this through the Gut-Brain Paradox Program.

But guess what else can help reduce apoptosis, particularly in your neurons? You guessed it—certain LPSs, which also enhance antioxidant activity in neurons, leading to fewer reactive oxygen species (ROSs).[16] So, let me make sure this is clear—when the immune system is familiar with LPSs, it keeps your neurons healthier, which helps prevent apoptosis. In addition, when your immune system is familiar with LPSs, it is less likely to attack the extracellular mitochondria stemming from any apoptosis that does occur. It certainly seems like it is about time to reacquaint our immune systems with LPSs, and with bacteria in general!

Gram-Positive Versus Gram-Negative Membranes

LPSs are part of the outer membrane of gram-negative bacteria. Have you ever asked yourself, *What about gram-positive bacteria? Don't they have an outer membrane, too?* I'm glad you asked.

In fact, the main difference between gram-positive and gram-negative bacteria is the thickness of that outer membrane. In the late 1800s, a bacteriologist named Christian Gram developed a test to determine whether bacteria had thick or thin membranes. Those with thick membranes are referred to as gram-positive, and those with thin membranes are considered gram-negative.

While the thin membranes of gram-negative bacteria contain LPSs, the thicker membranes of gram-positive bacteria contain exopolysaccharides (EPSs). I haven't written about EPSs before because they are not as potent as LPSs in terms of triggering an immune response. However, it appears that low doses of EPSs, like LPSs, can help train your immune system to no longer attack anything that looks or acts or smells (just kidding) like bacteria.

Like LPSs, EPSs can help improve overall immunity.[17] One study that looked at the immune-boosting effects of soy milk that was fermented with EPSs found that it had a strong ability to modulate immunity and could be used as an immunomodulatory functional food.[18] Side note, while soybeans do contain lectins, fermenting them drastically cuts down on their lectin content. So, I agree that fermented soy milk is a great idea.

In addition, EPSs can help maintain the gut lining and influence the production of cytokines, both of which reduce levels of inflammation. EPSs also help control the production of immune cells called lymphocytes and natural killer (NK) cells, which, as their name suggests, destroy infected or diseased cells, such as cancer cells.[19] On top of helping to prevent cancer, this helps cut down on the apoptosis that would normally occur when those diseased cells died.

In addition to being good for you, EPSs are good for your gut buddies and for your inner terrain. EPSs help protect friendly gut buddies from pathogens[20] and from antimicrobial peptides.[21] At the same time, they have their own antimicrobial effects against pathogens and can prevent bad bugs from proliferating.[22]

While I was never quite as concerned about EPSs as I was about LPSs, I am now ready to wholeheartedly embrace them, especially as a healthy part of gut-friendly fermented foods.

THE WHOLE GRAIN/BEANS LPS PARADOX

With all of this in mind, it has become clear to me that many of the plant foods with health benefits, including those used in plant-based medicine, aren't actually providing all of those benefits in and of themselves. They are also acting as a delivery device for the plant's microbiome, which provides additional benefits. This microbiome of course includes some living bacteria (probiotics), dead bacteria, postbiotics, and LPSs.

I wrote in my last book, *Gut Check*, about how dead bacteria actually include important messages for your gut buddies. In one study, supplementing with dead bacteria strengthened the gut lining and reduced inflammation more effectively than supplementing with living bacteria.[23] I argued then that the only problem with dead bacteria was those pesky LPSs that caused inflammation. I didn't realize that this was because our immune systems have not been trained to recognize them!

Now we see that consuming a plant's microbiome is a great way of getting a nice low dose of LPSs—and that, in fact, this may be why that plant is so good for us to begin with. Even common foods like olives provide bacterial benefits that we never previously considered. There are bacteria that form biofilms on olives, and when you ferment them into table olives, this bug is ingested and modulates the microbiome in powerful ways. Fermented pickles contain the same bug![24]

In addition, many herbs are well known for their ability to boost the immune system, but this may actually have little to do with the herbs themselves and everything to do with their microbiomes! A screening of over 2,500 natural plants with known immune-boosting effects showed that their microbiomes contained high numbers of gram-negative bacteria, which produce LPSs.[25]

In particular, one of the LPSs that helps make those herbs so healthy is from the bacteria *Pantoea agglomerans*. Learning about this LPS and what it can do for our immune systems has made me completely rethink one of my most controversial and strongly held stances.

I have been criticized quite consistently over the years for saying that whole grains are not actually good for you, despite the fact that they are a part of the notoriously healthy Mediterranean diet and that some studies do show that whole grains have health benefits.[26] I have been arguing for years that the Mediterranean diet is healthy *despite* those whole grains, not because of them.

My problem with whole grains is, in a word, lectins, those plant defense mechanisms that can help cause leaky gut, leading to neuroinflammation. In particular, one lectin that is present in wheat germ, called wheat germ agglutinin (WGA), binds to sialic acid, which is part of the coating on many of our cellular surfaces. These include the BBB and the myelin sheath that encases and insulates your nerves. This coating itself is called the glycocalyx.

When it binds to sialic acid, WGA acts as a splinter on the glycocalyx. The problem is, WGA is small and can leak through the gut barrier even if you don't have leaky gut.[27] WGA can also penetrate the BBB,[28] leading to inflammation in the brain and the rest of the nervous system.

Back to *Pantoea*. As I was reading about LPSs, I came across this gut buddy. Its LPSs are present in many of the herbs mentioned above, as well as whole grains. When grains are refined, the hull is removed, and so is the microbiome. No LPSs for you. It turns out that LPSs from *Pantoea* have many health benefits and can help remodel the immune system in incredibly powerful ways.[29]

When LPSs from *Pantoea* are placed under the tongue, they are able to train the immune system and prevent immune-related disease. And oral administration of these LPSs reduces inflammation overall. It also lowers low-density lipoprotein (LPL) cholesterol, helps decrease body weight, improves glucose tolerance, and reduces atherosclerotic plaque deposits.[30,31]

Pantoea LPSs are so powerful that there is a Japanese immune-boosting herbal medicine called Juzen-taiho-to (JTT) that includes ten medicinal herbs and their microbiomes, which contain this gut buddy. JTT is used in East Asia to boost the immune system, and studying this formula reveals that its effects stem at least in great part from these LPSs.[32] In addition to many other immune-boosting effects, JTT improves antiviral cellular immunity in tumors,[33] helps reduce inner ear inflammation in children,[34] and supports the growth of new microglial cells in the brain.[35]

It's clear that we want to consume some of these LPSs—ideally in consistent low doses. So, I couldn't help but laugh when I learned that they are present in whole grains, as well as beans,[36] another so-called healthy food that I have also famously (or infamously) railed against because of their lectins! It's fascinating to think that the benefits of whole grains and beans might come from the LPSs boosting the immune system and not from the foods themselves. An absolutely mind-blowing (quite literally) paradox.

As a side note, do you remember earlier when I mentioned that there is a flavonoid in hops that reduces levels of pro-inflammatory microbes? It turns out that our new favorite LPSs from *Pantoea* are present in many types of hops! Maybe that's the real reason that these hops are good for you and why our ancestors were so wise to include them in their beer.[37]

Our ancestors were so wise, in fact, that they usually fermented their grains and beans. This helped accomplish a few things. For one, during the fermentation process, bacteria eat the vast majority of lectins, making these foods much safer for your gut and therefore for your brain. Fermenting foods containing lectins is one of the best ways to reduce their lectin content so that you can safely eat them without damaging your gut wall in the first place.

Further, fermented foods contain dead bacteria, LPSs, and the postbiotics that are produced during the fermentation process. So, eating fermented whole grains and beans makes them safer while ensuring that we are getting all of the benefits of bacteria, including more familiarity with LPSs. No wonder, then, that a diet high in fermented foods decreases inflammation.[38]

Discovering this new information about LPSs and particularly their connection to whole grains and beans was like fitting in that last satisfying piece of the puzzle. I am truly excited to help you leverage this knowledge in the Gut-Brain Paradox Program. But before you reach for that fermented sourdough bread at the artisanal bakery here in America, buyer beware! That wheat was almost certainly sprayed with glyphosate (Roundup), one of the best gut disrupters ever invented. But, if you are reading this book in Europe or most of Asia and Africa, go ahead and enjoy that LPS-bearing bread.

In case you are wondering why I don't endorse fermented organic breads from the US, it's because of my years of following my patients' experiences. Most of my patients with leaky gut and autoimmune diseases reverse their diseases and leaky gut within a year by following the "yes" and "no" lists of my program, which exclude grains like wheat, rye, barley, corn, oats, and brown rice (to name a few). Once in remission, many of them then travel to Europe or Asia and "indulge" in local foods such as bread, pasta, and pizza, and to their surprise, they don't react to these foods! Needless to say, they are delighted, but when they return to the US thinking that they are "cured," they begin eating the equivalent US foods (even organic varieties) and, within weeks, reactivate their autoimmune diseases and/or leaky gut issues. It's happened so many times that it's just not worth it to give you the go-ahead, as much as I'd like to.

CHAPTER 5

YOUR MICROBIOME PREDATES YOU

One of the first things your mother ever gave to you was the foundation of your inner terrain. In fact, she gave this gift to you back when you were still in her womb. Yes, the exact mix of microbes in your gut was determined to a great extent before you were even born. But your mother's own microbes have had an impact on your development that is even further reaching than that.

For one thing, the microbes from your mother that you were exposed to in utero played a big role in shaping the makeup of your own microbiome, your immune system, and therefore your general health. Of course, they also had a dramatic impact on your brain. I'm not just talking about the overall health of your brain and your levels of neuroinflammation, either. No, the microbes that you were exposed to in utero also impacted your behavior, your personality, or the likelihood of you ever struggling with addiction or mental health or being neurodiverse.

Let me be clear. I am in no way trying to "blame" or shame mothers for their children's neurological or behavioral issues, differences, or challenges. Quite the opposite, in fact! While so many parents fear that they've done something "wrong" when they see their children struggle—and I can relate to this myself—I believe the information in this chapter can help liberate all of us from these misapprehensions.

The simple truth is that so much of our brain's fate is actually in the "hands" of our microbes. That means that brain-related issues or neurological differences are not any person's "fault." Even better, however, is the fact that those fates are not written in stone. We can change our microbiomes and change our brains, and therefore reverse these issues. And that is exactly what I aim to help you accomplish.

WINDOWS OF DEVELOPMENT

While our brains certainly have the ability to grow and change throughout our lifetimes, there are three distinct periods during which neuronal development is at its most rapid. It is therefore essential to ensure the brain has everything it needs to grow and become as healthy and high-functioning as possible during these times. These are: pregnancy (in utero), birth, and the first year of life.

Pregnancy

During pregnancy, the mother's gut biome impacts the fetus in numerous ways, primarily through postbiotic signals produced in her gut. Postbiotics produced in the mother's microbiome are able to pass from her gut to the placenta and then to the fetus. Pretty amazing! For instance, SCFAs produced in a mother's intestines can cross the placenta and impact the fetus's brain development.[1] They also influence overall nervous system development and function.[2]

For example, during pregnancy, the mother's microbiome plays a role in what's called fetal thalamocortical axonogenesis, which is the axonal branching that connects the fetus's thalamus to the cortical areas of the brain. As you likely recall from biology class, axons are long projections of nerve cells that conduct electrical impulses. The mother's gut buddies send metabolites to the fetus's neurons that support this axon outgrowth, connecting and allowing messages to be sent between different parts of the brain.

In studies on mice, eliminating the maternal microbiome impairs fetal thalamocortical axonogenesis, which leads to neurobehavioral changes in offspring that last until adulthood. Well, you may be thinking, *That was in mice. What about humans?* Simply put, it's not a good idea to replicate this study on humans. The good news, however, is that restoring the mother's microbiome or supplementing with specific metabolites eliminates these changes in her future offspring.[3]

Of course, with the mother's inner terrain playing such a big role in fetal development, anything that changes a pregnant mother's microbiome can also lead to changes in the developing fetal brain. For instance, a study looking at one million births in Finland revealed that children who had been exposed to antibiotics in utero had an increased risk of developing sleep disorders, attention-deficit/hyperactivity disorder (ADHD), conduct disorders, mood and anxiety disorders, and other behavioral and emotional disorders.[4] Yikes. Again, these things are not your fault. They are the fault of a disrupted terrain (or of a doctor who is too readily handing out antibiotics . . .).

Further, we already know that drug and alcohol use during pregnancy is harmful to the growing fetus, but why? One reason is that drugs and alcohol cause changes to the mother's metabolites! For instance, one study looked at levels of glutamate, glutamine, and serotonin (which you recall are all produced in the gut) in mothers who were exposed to drugs and/or alcohol during their first trimester of pregnancy versus those who were not. There were significant differences. The mothers who used alcohol and drugs during their first trimester had higher glutamate levels and lower glutamine levels than the mothers who did not, while mothers who used alcohol during their first trimester also had reduced levels of serotonin compared to mothers who did not. These changes directly impact the fetus's growing brain.[5] Moreover, these effects stem from changes in—you guessed it—the microbiome![6]

A pregnant mother's microbiome may also become imbalanced because of diet and/or obesity. Exposure to a maternal high-fat diet that triggers obesity can induce long-term cognitive effects that span generations.[7] Infants who are born to mothers with obesity have distinct microbiome profiles compared to infants born to mothers who are at a normal weight. And a study of 778 children ranging from seven to fourteen in China showed that maternal obesity is strongly associated with a reduction in children's cognition.[8]

Meanwhile, female mice with obesity give birth to offspring with synaptic impairments and microglial defects in their brains. Amazingly, these impairments are alleviated by feeding either the pregnant mothers *or* the offspring a high-fiber diet. This shows exactly how reversible these brain changes can be. Feed the gut buddies what they want, and they will repay you by boosting your brain. In addition, treatment with a mix of acetate and propionate—two SCFAs normally produced in the gut and that serve as building blocks for butyrate production—significantly improves these mice's cognition.[9]

Birth

At the time of birth, the baby's gut is "colonized" as it travels through the birth canal and is exposed to the mother's vaginal microbiome.[10] As such, there are marked differences in the guts of babies who are born via cesarean section (C-section) and those who are born vaginally. Again, I am not sharing any of this with the intention of blaming or shaming mothers who give birth via C-section. C-sections can be and often are lifesaving for both the mother and the soon-to-be-delivered child. It is also possible to compensate for the ways that different modes of delivery impact the child's microbiome and therefore brain health. However, in order to make these compensations, we need the information about what those changes are.

During a C-section, the infant is not exposed to, and therefore not colonized by, the mother's vaginal microbiome. Interestingly, the microbiomes of newborns who are born via C-section are colonized instead by bacteria that are present on the mother's skin and in the environment they are first exposed to at birth.[11] This leads to a different mix of microbes in the newborn's gut. Babies who are born via C-section tend to have reduced microbial diversity and richness of their inner terrains.[12] More worrisome, they are often missing specific bacterial species that are key for brain development and contain more pathogens instead.[13,14]

There are also differences in the microbiomes of babies born via planned versus emergency C-sections. In the case of an emergency C-section, the infant is likely to be partially exposed to the birth canal—and therefore its microbiome—during the early stages of labor. As such, babies born via emergency C-section have microbiomes that are more similar to those born vaginally than those born via planned C-section.[15]

But there's an additional way that the method of birth impacts the infant's microbiome. When a mother is prepared to undergo a C-section, she is routinely given antibiotics to prevent infection. This not only impacts her microbiome, and therefore that of her infant, but it also reduces the microbial diversity and richness in the mother's breast milk microbiome.[16] (Yes, human breast milk is loaded with bacteria. More on this in a moment.)

With these changes to the infant microbiome, human studies have revealed a link between babies who are born via C-section and their cognitive ability later on. One study showed that babies born via C-section were more likely to experience a delay in cognitive development at the age of nine months compared to babies born vaginally. But before you panic, these babies "caught up" and no longer experienced delays by the age of three.[17] Another study showed a small but significant difference in school performance between babies born via C-section and those who were born vaginally.[18]

Children born via C-section are also more likely than those born vaginally to be diagnosed with ADHD. This was confirmed by a 2019 review of sixty-one different studies looking at more than twenty million deliveries,[19] and was again confirmed by a new study in 2023.[20] ADHD is related to abnormal dopamine nerve transmission. As we know, dopamine is produced in the gut. Indeed, patients with ADHD have altered microbiomes with a higher relative abundance of the gut buddies that produce dopamine.[21] The good news is that targeted probiotics have been shown to successfully treat ADHD in children.[22]

In mice, those born via C-section experience increased neuronal death, leading to a decrease in certain types of neurons. They also experience changes in infant vocalization compared to mice who are born vaginally.[23] (As an interesting aside, mice "talk," but we can't hear them, as their voices are in the ultrasound range. Instead of saying "quiet as a mouse," perhaps we should say "deaf as a human.") Mice born via C-section are also more likely to experience anxiety-like symptoms and reduced cognitive ability than mice who are born vaginally. However, have no fear—these changes, like so many of these ones we've discussed, can successfully be reversed by the administration of probiotics![24]

The First Year

After birth, the baby's first year of life is an incredibly important period for neurological development, as both the gut and the brain are at their most malleable during this time. Dysbiosis in the gut within the first year of life may negatively impact neurodevelopment and therefore cognitive function, and these changes can be long-lasting.[25,26] For example, mice who are born with normal microbiomes but are treated with antibiotics within the first few weeks after birth experience cognitive and behavioral changes later in life.[27] This is due to alterations in brain development during this important window.

There is also a fascinating and very telling connection between the human infant's microbiome composition and temperament. One study looked at the microbiome makeup of newborns during their first three weeks of life and then assessed their temperament at the age of twelve months. A diverse microbiome with an abundance of specific gut bugs (*Bifidobacterium* and *Lachnospiraceae*) during the first three weeks of life was associated with an outgoing, friendly temperament at twelve months. Meanwhile, an abundance of another bacteria, *Klebsiella*, during the first three weeks of life was negatively associated with a friendly, outgoing temperament at twelve months.[28] If you've ever wondered why your baby seemed so crabby, look no further! It may just be their gut bugs calling the shots!

Ideally, the newborn's gut and neurodevelopment are modulated immediately after birth and throughout the first year by breast milk, which contains bacteria, as well as their postbiotic metabolites and a generous dose of oligosaccharides. (More on them in a second.) There's that breast milk microbiome I mentioned earlier, which plays a big role in colonizing the infant gut. This lays the foundation of the gut-brain axis and therefore impacts brain function throughout life.[29] Once again, there are ways mothers can compensate if they cannot or choose not to breastfeed. But in general, breast milk is formulated to establish and nourish a healthy terrain in the infant.

Many human studies reveal the differences between the inner terrains of formula-fed versus breastfed infants.[30,31,32,33] Of course, some of these differences stem from breastfed infants ingesting the actual bacteria in breast milk. In addition, however, one of the primary ingredients in breast milk is there not to feed the infant itself, but to feed the infant's microbiome. After lactose and fat, the third most abundant component in breast milk is human milk oligosaccharides (HMOs). However, infants cannot digest HMOs. So, what are they doing in there? I'll give you one guess who *can* digest it. That's right—the infant's little baby gut buddies, of course.

HMOs are prebiotics—aka food for bacteria. Mother Nature designed breast milk to feed not just the baby, but also to feed the baby microbiome and nurture its own growth and development.[34] There are more than two hundred separate types of HMOs in breast milk, which feed and nourish different strains and species of bacteria.[35]

However, it's important to note that the mother's diet also helps shape her breast milk microbiome. Mothers who consume a diet that is high in saturated fat produce breast milk with lower-than-normal levels of HMOs. This can provide inadequate nutrition for the infant's gut buddies, leading to an imbalanced terrain.[36]

In addition to HMOs, there is another substance in breast milk that makes it especially beneficial for the infant's brain. For years, I've been writing about milk fat globule membranes (MFGMs), which surround the fats in milk and make them soluble. In the past, I've been a vocal fan of MFGMs because of their positive effects on mitochondria. By boosting mitochondrial health, MFGMs aid in weight loss and preventing insulin resistance, which reduces diabetes risk.[37,38]

But MFGMs in breast milk also benefit infant neurological development, and they do this by making changes to the microbiome.[39] When female rats are fed a high-fat diet that triggers obesity and changes the makeup of their microbiomes, their offspring experience significant delays in neurogenesis and in their neurological reflexes. When these female rats are given supplements of MFGM during pregnancy and lactation, their offspring have microbiomes that are more diverse and have a greater abundance of friendly gut buddies with fewer pro-inflammatory bacteria. This leads to fewer LPSs, a decrease in microglia activation, and reduced neuroinflammation. The result? A reversal of the negative neurological effects of the mother's high-fat diet.[40] Powerful stuff, those MFGMs.

The good news here is that it's possible to include MFGMs and/ or its components in infant formula. Multiple human studies show that infant formula containing MFGMs provides many of the same benefits as breast milk and positively affects infant neurocognitive development.[41,42,43]

AUTISM AND THE INFANT TERRAIN

According to the Centers for Disease Control, as of 2020, one in thirty-six children was diagnosed with autism. The current number is likely far higher, as it has been going up consistently and dramatically year after year. Unfortunately, this does not surprise me, and I assure you that the increase in autism rates coinciding with massive changes to the way we eat and how we grow our food is not a coincidence.

Although some people seem to not want to believe that there is a direct connection between autism and the gut, the evidence is truly overwhelming. In fact, as I was doing the final edits on this book, a study was published looking at stool samples from 1,600 children ranging in age from one to thirteen. The results showed that children with autism have distinct biological markers in their stool, including unique traces of gut bacteria, fungi, and viruses.[44]

Indeed, we have known for a long time that children with autism also experience gastrointestinal issues, and that autism in general is associated with leaky gut and alterations to the inner terrain.[45] Further, when gut microbiota from humans with autism are transplanted into germ-free mice, it induces hallmark behaviors associated with autism. Treating those mice with metabolites from a healthy microbiome modulates the activity of neurons and improves those behaviors.[46,47]

In addition, autism is associated with mitochondrial impairment.[48] This makes sense due to the close connection between the inner terrain and mitochondria. If our mitochondria do not get the right signals from the gut, they cannot function properly and are more likely to die via apoptosis. All of this triggers neuroinflammation.

Many human studies on children with autism show that fecal transplants significantly improve their symptoms. One study involved giving forty children with autism between the ages of three and seventeen fecal transplants, all from the same donor. Each of these children showed significant improvements in their gastrointestinal symptoms, which included abdominal pain, constipation or diarrhea, and reflux. In addition, the children's autism symptoms were improved. Scores on tests that assess mood, behavior, emotion, and language were significantly improved eight weeks after the fecal transplant.[49]

In another study, children with autism who received a fecal transplant experienced improvements in their gastrointestinal symptoms that lasted for at least two years. And a professional evaluator found a 45 percent reduction in these children's symptoms related to language, social interaction, and behavior.[50] Pretty remarkable.

Interestingly, we are now seeing that many of the changes to the internal terrain that can lead to or contribute to autism actually occur before birth and are the result of changes to the mother's microbiome. For instance, a study of all children born in Denmark over the course of twenty-five years showed that mothers who were admitted to the hospital during their first or second trimesters due to an infection had a higher chance of giving birth to a child with autism than those who had not been hospitalized with an infection.[51]

Further, male children with autism were more than three times as likely to have been exposed to aspartame (a common artificial sugar in diet sodas and other products) daily either in the womb or during breastfeeding.[52] I've written before about the disastrous effects that aspartame can have on the microbiome. It kills normal bacteria and promotes inflammation.[53] It also negatively impacts the brain in other ways that appear to be heritable.

When mice are fed aspartame daily for sixteen weeks at doses that are a mere 7 to 15 percent of the FDA's recommended maximum daily intake, they experience significant spatial learning and memory deficits. Even more shocking, those cognitive deficits are passed on to their descendants along the paternal lineage![54] I urge you to give your descendants the gift of you not ingesting any aspartame. There is truly no "safe" dose.

While it appears that anything that harms the mother's inner terrain can also increase the chances of her children having autism, the reverse is also true. Enhancing a mother's microbiome decreases her children's risk of autism. For example, higher-than-average vitamin D levels in the mother are associated with a reduced risk of both autism and ADHD in her children.[55] And vitamin D supplementation improves symptoms in children with autism.[56]

What does vitamin D have to do with the gut? Quite a lot, actually. It turns out that vitamin D has a bidirectional relationship with the microbiome.[57] Vitamin D affects your mix of gut buddies, and you need the right mix of gut buddies to synthesize vitamin D for you.[58]

One of the main benefits of vitamin D is that it increases the abundance of our favorite gut buddy, *Akkermansia*, along with two other major butyrate producers (*Faecalibacterium* and *Coprococcus*).[59] This means that vitamin D drastically increases your microbiome's butyrate production. And, as you recall, butyrate helps support the gut wall as well as the BBB. This is your sign to go out and get some sun and to supplement with vitamin D_3.

Indeed, butyrate supplementation enhances mitochondrial function in children with autism.[60] In fact, pretty much any intervention that helps heal the gut—including simply taking probiotics—consistently improves autism symptoms.[61]

* * *

While the gut and brain may be at their most malleable before, during, and for the year after birth, they both remain changeable throughout the course of our lives. It's never too late to improve the health of your inner terrain to create positive changes in your brain—and in the brains of your children and your descendants who follow.

CHAPTER 6

THE ADDICTION MICROBIOME

The evidence has been mounting for years that people struggling with addictions, whether they are to alcohol, cigarettes, or even opioids and other painkillers, have altered microbiomes. Until now, it has generally been assumed that this was a simple case of cause and effect—those addictive substances changed the makeup of the inner terrain. End of story, nothing to see here, folks. Well, it turns out that, like most things, it's not quite that simple.

It's no exaggeration to say that the information I uncovered while doing the research for this chapter absolutely shocked me. In fact, it completely changed the way I now think about various addictions. As I'll discuss in great detail, it appears that these addictions are not driven by the individual who is seeking an addictive substance, but rather by the individual's microbiome.

This may sound controversial, but once you get used to the idea, it makes perfectly good sense. We already know that our gut buddies send postbiotic messages to our brains, telling them what foods to eat—the ones that allow them to grow and thrive. As I keep saying, your gut buddies are the ones calling the shots. So why would this be any different when it comes to other substances that you ingest, inject, or inhale? As you're about to see, it isn't.

The simple truth that I've come to accept is that our gut bugs are going to get what they want one way or another. They are incredibly wise and crafty and have evolved over billions of years to survive. There's no sense in fighting them—and fighting them wouldn't do you any good, either. After all, you quite literally can't live without them. There is even evidence that they made you![1]

No, this doesn't mean that you should satisfy a drug-seeking microbiome with drugs any more than you should make a sugar-seeking microbiome happy by eating more sugar. It does mean, however, that you should aim to heal your inner terrain. With a balanced microbiome, you will begin to crave the things that are good for *you* instead of those that are addictive and/or harmful and feed the bad bugs in your gut rather than your helpful gut buddies. This is how you can win both the battle and the war, and it's exactly what I'm going to help you do on the Gut-Brain Paradox Program.

You may already be asking yourself, *If our gut bugs are calling the shots, then why do some people develop addictions while others don't?* This is the multimillion-dollar question. Like everything you ingest, addictive substances and your microbiome have a bidirectional relationship. They each have an impact on the other. In many cases, people who develop addictions already have altered microbiomes beforehand, which makes them more susceptible to becoming addicted. Yes, really.

But whether or not someone suffering from addiction had an altered terrain before becoming addicted to that substance, that substance itself creates harmful changes to the microbiome that drive substance-seeking behaviors. Read that again. Although I'll go into great detail on specific substances, each of them fosters specific bacteria that can grow and thrive among these substances, which tend to be the bad bugs, and kills off the helpful gut buddies that cannot. This leads to dysbiosis and leaky gut, which triggers the release of inflammatory cytokines. Worst of all, this creates a microbiome made primarily of bacteria that rely on that substance in order to thrive, so they call for more of it.

Back to those cytokines for a moment. They increase pain. Inflammation, and particularly neuroinflammation, hurts.[2] As you'll see, these sensations of pain from inflammation increase substance-seeking behavior and even drive drug tolerance, which is when you need more of a substance to experience the same effects. But this is exactly what the bad bugs wanted to begin with. The pain is there specifically to tell you to ingest more of that thing that started this vicious cycle. When you eventually give in and ingest that substance because you can't tolerate the pain any longer, the bad bugs get what they want. This crowds the good gut buddies out even more and allows the bad guys to proliferate and thrive.

Then what happens? More leaky gut, more inflammation—aka more pain—and eventually seeking more of that substance. This vicious cycle repeats itself again and again and is at the root of addiction. And it explains exactly why certain addictions can seem so darn difficult, if not impossible, to break.

I truly hope that the information in this chapter liberates many families of those suffering from addiction and those suffering themselves from shame and stigma. While it is often said that addiction is a disease—and I agree that this is true—I would go a step further to argue that it is much more than just a psychological disease, but a physiological one that is stemming from the gut. Or perhaps even a symptom of dysbiosis and leaky gut.

Again, the good news is that the gut can be healed. I have witnessed truly amazing recoveries that should offer hope to anyone who is suffering from addiction or suffering while watching a loved one suffer. I promise you it is possible to break the cycle. First, let's take a closer look at how that cycle plays out with different addictive substances.

THE ALCOHOL-SEEKING MICROBIOME

We know that overconsumption of alcohol leads to changes to the brain and to the microbiome.[3] One study compared the microbiomes of patients who overconsumed alcohol for more than ten years to controls who had consumed very little or no alcohol. The overconsumers had a higher relative abundance of certain pro-inflammatory bacteria and a lower relative abundance of helpful gut buddies, including certain species of SCFA-producers. As a result, they had lower concentrations of butyrate and disruptions to their epithelial cell junctions (aka leaky gut).[4] The study concluded simply that overconsumption of alcohol skews the microbiome in a pro-inflammatory direction.

This is no surprise. It has consistently been shown that over-consumption of alcohol causes dysbiosis, leaky gut, and immune system activation, leading to inflammation.[5] These changes certainly contribute to, if not directly cause, many of the physiological effects of alcoholism, such as inflammation of the liver and alcoholic liver disease (ALD).[6]

But don't forget who processes the alcohol that you digest in the first place—your gut buddies, of course. Again, alcohol and your gut buddies have a bidirectional relationship. (Are you sick of that term—bidirectional relationship—yet? You'll be seeing it a lot more in this chapter.) Alcohol consumption influences microbiome composition, and the makeup of the microbiome influences how alcohol is metabolized and therefore affects the body. This may explain why some alcoholics develop ALD and others do not. It all comes down to the makeup of the inner terrain.

For example, one study showed that a subset of those suffering from alcohol addiction had persistent leaky gut and alterations to their gut biomes. This led to a pervasive inflammatory state. These inflamed patients were more likely to develop endotoxemia and ALD compared to those suffering from alcohol addiction whose leaky guts weren't as persistent.[7]

The state of the inner terrain before alcohol has a chance to affect it may also help explain why some people can drink large amounts of alcohol without becoming addicted while others cannot. It is not simply a case of alcohol changing the microbiome and the microbiome causing you to seek more alcohol (although that happens, too). As studies have noted, gut dysbiosis is associated with metabolic changes that affect the behavioral and neurobiological processes involved in alcohol addiction.[8] For example, a study of binge drinkers in their teens and early twenties showed that binge drinking and specifically cravings for alcohol were associated with distinct microbiome alterations.[9]

Further, a prospective cohort study with a thirty-year follow-up looked at microbial profiles of men who were later treated for alcoholism and compared them to the microbial profiles of men who were never treated for alcoholism. There were significant differences between the two groups when it came to their circulating levels of sixty-four different bacterial-derived metabolites. In particular, the men who were later treated for alcoholism had lower levels of asparagine, an important amino acid that is produced in the gut,[10] and serotonin, the neurotransmitter that you read earlier is made in the gut from tryptophan. This means that the men who ended up suffering from alcohol addiction later on were having different messages sent from their guts to their brains than men who never suffered from addiction—and this was long before their alcoholism took hold.[11]

Another compelling study showed that patients who suffered from alcohol addiction and had leaky guts and an altered terrain continued to experience depression, anxiety, and alcohol cravings after going through an alcohol detoxification program. This is such a common experience for those suffering from alcoholism—repeated periods of detoxification with eventual relapses—and it can be so painful and demoralizing. We now have evidence that leaky gut makes it harder to break the cycle of addiction.

In fact, patients whose leaky guts healed during the period of detoxification had an easier time staying away from alcohol and experienced greater improvements in their psychological symptoms compared to those whose leaky guts persisted![12] If this doesn't illustrate the importance of a healthy microbiome and gut lining, I'm not sure that anything could.

As is also the case in the other addictions we'll discuss, alcohol abuse creates alterations to the microbiome and worsens leaky gut, and each of these problems compounds the other. Together, this dramatically ramps up inflammation and contributes to disease progression.[13] All of this inflammation also creates direct changes in the brain. Those suffering from alcohol addiction have brain region–specific increases in microglial markers. This leads to neurodegeneration as well as an increased desire for alcohol.[14]

Like I said, one way or another, those gut bugs are going to get what they want. The answer, therefore, is not to try to "sweat it out" or go "cold turkey" to try to give up alcohol. The only lasting solution is to change the makeup of your gut buddies themselves. As long as the bad guys are still around down there and you are depriving them of what they want, the messages from them to your brain saying to "give in" will keep getting louder and more persistent.[15,16]

Indeed, studies looking at treating alcoholism via the microbiome are very promising. In a well-designed study, patients suffering from alcohol addiction received either a fecal microbiome transplant or a placebo: 90 percent of the group that received the fecal microbiome transplant and 30 percent of the placebo group experienced significant decreases in alcohol cravings and improvements in cognition and quality of life. The group that received the transplant also saw an increase in microbiome diversity and SCFAs, while the placebo group did not.[17]

Luckily, we can transform the microbiome and shift it away from its inflammatory, alcohol-seeking state and see similar results without a microbial transplant by following the Gut-Brain Paradox Program.

Jack and his live-in girlfriend flew to see me a few months after he had suffered a massive, life-threatening heart attack at the tender age of thirty-two. He survived thanks to the modern intensive care he received, but much of his heart muscle was permanently damaged. And he didn't want a transplant or an implantable defibrillator.

Jack also didn't want his health to impact his career as a comedian and the lifestyle that went along with it. Jack did not view himself as an alcoholic, but his consumption of alcohol was excessive, to say the least. Jack had been warned that his high level of alcohol consumption was at the root of his heart attack and that continued use would further damage his heart muscle. But Jack felt unable to make a change. He came to me for help with his heart, not his alcohol consumption.

Jack's blood work assured both him and me that cholesterol, which is often blamed for heart disease, wasn't his problem. But his inflammation markers were very high across the board. And his intestinal permeability (leaky gut) and LPS scores were very high, as well. And his microbiome diversity? Forget about it! He had no mucus-producing bacteria, no SCFA-producing bacteria, and mostly bad guys running amuck and calling the shots. It was all there in black and white for Jack to see.

I put him on some heart muscle stimulating supplements that I developed when I was chief of heart surgery at Loma Linda University Medical Center to keep patients like Jack off my heart transplant list. Most important, I changed his diet to the Gut-Brain Paradox Program and supplemented him to heal his gut and create a more hospitable terrain. And get this: I allowed him to have a glass of red wine per day, but no other alcohol. He agreed.

There were a lot of phone calls and follow-up blood work and echocardiograms. The leaky gut slowly healed, the LPSs stopped getting into his blood, the inflammation subsided, and Jack's heart function stabilized. No, he wasn't perfect in terms of his alcohol consumption, but he now found it unpleasant when he drank too much. He told me that it was like there was an inner voice telling him how stupid it was. And you know who that inner voice was? His newly in-charge gut buddies, of course.

Jack and I recently celebrated his good health and fourteen years as my patient. He's gotten married to his girlfriend, is still working as a comedian, and has a completely new support team for a microbiome. He's still got his same heart, but he's a whole new person with a whole new terrain that enjoys just a glass of red wine and asks for no more.

THE CIGARETTE-SEEKING MICROBIOME

As with alcohol, the use of tobacco (in cigarettes or any other tobacco products) creates changes to the microbiome[18,19,20] by raising the pH levels of the environment, triggering chronic inflammation, and inducing oxidative stress.[21,22,23] In particular, these changes kill off an important group of gut buddies called *Bifidobacterium*. Numerous studies have shown that smokers have a significantly lower relative abundance of *Bifidobacterium* compared to nonsmokers.[24,25] The good news is that even short periods of smoking cessation can help restore *Bifidobacterium* levels, to a degree.[26]

But the relationship between *Bifidobacterium* and smoking is more complex than a simple case of tobacco killing off this species. It turns out that children who had a lower abundance of this bacteria to begin with were more likely to start smoking at a younger age compared to people with normal levels of this gut bug.[27] How can this be?

Of course, it goes back to the signals being sent from the gut to the brain. *Bifidobacterium* is so important in part because it sends signals to the brain that cause it to secrete higher levels of dopamine, the neurotransmitter associated with the brain's reward and pleasure functions.[28] Guess what else triggers the release of dopamine? Nicotine—the main chemical in tobacco.

It appears that people who already have low dopamine levels due to dysbiosis and a lack of *Bifidobacterium* are more likely to become addicted to the dopamine hit they get from smoking. And, of course, the more you smoke, the fewer *Bifidobacterium* and therefore the less dopamine you have! This makes cravings for cigarettes stronger, causes smoking to feel even better, and makes it far more difficult to quit.[29]

There is a direct connection between tobacco and other species of bacteria, too. For example, smoking decreases the abundance of *Actinobacteria*, which is the single most prolific source of bioactive secondary metabolites in the gut![30] By sending all those postbiotic messages, these bugs clearly have a tremendous amount of power over your brain. Having a low abundance of *Actinobacteria* causes smokers to increase the number of cigarettes they smoke per day—which, of course, kills off even more of this species. This is another vicious cycle contributing to addiction.[31,32]

What's more, it is well established that weight gain usually follows smoking cessation. While we know that nicotine suppresses appetite, it shouldn't surprise you by now that smoking cessation causes overeating due to chronic changes in the microbiome that drive food-seeking behavior![33,34]

THE OPIOID GUT EPIDEMIC

We've known since the 1960s that some of the most common painkillers in the world cause dysbiosis. These are nonsteroidal anti-inflammatory drugs (NSAIDs), such as aspirin, ibuprofen, and naproxen. Studies on NSAIDs back in the 1960s were among the earliest documentation of what's called drug-induced dysbiosis. Yet their popularity remains. Upward of thirty million people take NSAIDs daily, but these drugs wreak havoc on the gastrointestinal system and create significant changes to the microbiome.[35]

NSAIDs shift the composition of the microbiome so that it is dominated by gram-negative rather than gram-positive bacteria. They also increase the relative abundance of specific pathogenic and pro-inflammatory bacteria such as *Enterococcus* and reduce the abundance of protective species, such as *Lachnospiraceae* and *Ruminococcaceae*, which both produce that all-important butyrate, which protects our brains and our gut walls.[36]

But, remember, you're presumably taking those NSAIDs because you're in pain. And what's a little dysbiosis in exchange for alleviating that pain, right? Not so fast. While popping an Advil may offer temporary relief from a headache, long-term NSAID use actually *increases* inflammation—and therefore pain.

One study showed that people who took NSAIDs for osteoarthritis ended up with more inflammation and poorer cartilage quality compared to a control group.[37] Well, of course they did! NSAIDs drive dysbiosis, which leads to inflammation, which leads to pain, which leads to more NSAIDs, which leads to more dysbiosis and more inflammation and more pain. Where does it end?

While I've been writing about the dangers of long-term NSAID use for a while, I have only recently discovered that other, more dangerous painkillers work in very much the same fashion. I'm talking about opioids, which include illegal drugs such as heroin, as well as prescription painkillers like oxycodone, Vicodin, codeine, morphine, fentanyl, and so on. Together, these drugs have been responsible for a steadily increasing number of overdose deaths for years, beginning back in the 1990s, when they started being increasingly prescribed for pain.

The opioid epidemic was officially declared a national public health emergency in the US in 2017. Many people and organizations have been blamed for this crisis, ranging from the pharmaceutical companies that market and sell these drugs to the doctors that prescribe them, and even the companies that advise pharmaceutical companies on how to drive sales. Others have blamed the government for failing to step in sooner. I'm pretty sure, however, that no one has thought to blame the bacteria living in our guts that are driving addictive behaviors based on their need for opioids—until now.

Like NSAIDs, opioids cause gut dysbiosis and significant gastrointestinal effects. This makes sense, since opioid receptors are highly expressed within the digestive tract. Perhaps this should have been the first clue telling us to look at the gut as a major player in the opioid crisis.

Indeed, morphine leads to opioid-induced dysbiosis and leaky gut after just *one day* of treatment. This particular brand of dysbiosis includes a reduction of our good friend *Akkermansia* and its fellow mucin-degrading gut buddy *Verrucomicrobia*.[38] This leads to the disruption of the gut barrier and resulting inflammation in the intestines, mesenteric lymph nodes, and remote organs.[39]

Opioids also activate microglial cells in the brain. Mice with opioid-induced dysbiosis have increased microglial cell body size in both the midbrain and dorsal horn of the spinal cord.[40] Yes, that translates to neuroinflammation—the alarm going off in the brain of someone who is suffering from addiction, telling them that they need more of this stuff, and they need it NOW.

In addition to creating widespread inflammation and pain, this opioid-induced dysbiosis also leads to a number of changes in how opioids work in the body. This, too, contributes to addiction.[41] Of course, your gut buddies are the ones processing those opioids for you. With opioid-induced dysbiosis, morphine is not recirculated in the bloodstream as it should be, and it begins to have a reduced efficacy. This is exactly how addiction starts—levels of pain go up due to increased inflammation, while the amount of relief that you get from the same dose of a drug goes down.[42]

Indeed, there is much evidence that the gut biome drives morphine tolerance, meaning the need for increased doses to get the same amount of relief. "Tolerance" may sound like a good thing, but I assure you that in this case, it is not. It indicates that your body and your opioid-seeking bad bacteria are accustomed to opioids and need a higher dose in order to feel relief from the same amount of pain.

The even bigger problem is that in this scenario, you're not experiencing the same amount of pain. Pain levels go up due to dysbiosis, leaky gut, and inflammation while the drug's pain-reduction efficacy goes down. This is all the result of an opioid microbiome working to keep you addicted.

Think about it this way—you have some bad bugs in you. Always have, always will. But with a balanced inner terrain, the gut buddies keep the bad bugs in check. They are not allowed to overgrow. Then opioids come along and kill off some of the good gut buddies, and suddenly, everything changes. The bad bugs in your gut may or may not love those opioids, but they can tolerate them better than some other gut buddies can. And they recognize right away that those gut buddies that have been keeping them in check all this time are susceptible to this substance.

The bad bugs have finally discovered their enemy's kryptonite. They rally together and say, "This is our chance! The more of this stuff we get, the more of those other bugs we'll get rid of. Soon, we'll be the ones in charge." They are smart and cunning, and they are able to perfectly set the stage so that you need more and more of the exact substance that kills off the good gut buddies, ultimately giving the bad bugs more power.

Indeed, studies show that morphine-tolerant mice have a distinct gut biome compared to mice who are not morphine tolerant—and this morphine-tolerant microbiome mirrors the microbiomes of human patients with substance use disorders. When these morphine-tolerant mice are treated with antibiotics, they are no longer morphine tolerant! This makes sense, since the bad bugs that were driving the addiction are wiped out.

However, when the mice have their microbiomes restored with the same gut bugs they had before, their morphine tolerance returns. Fascinatingly, when the mice are treated with probiotics, their morphine tolerance decreases. This shows that adding some gut buddies to the mix shifts the overall microbiome composition and helps restore inner harmony.[43]

But hold on a second. What is driving morphine tolerance to begin with? Our own human kryptonite, inflammation. Once again, there is a vicious cycle of opioid-induced dysbiosis triggering inflammation, which drives morphine tolerance while causing pain, all while making that dysbiosis worse and worse.

Opioids literally ramp up pain levels, leading patients to need higher doses. These higher doses increase pain even more by causing more dysbiosis and leaky gut and inflammation. And so on. Unfortunately, we know where this cycle too often ends.[44] Don't blame yourself if you cannot quit taking these drugs. It's not your fault.

Plus, there's even more. Opioid-induced dysbiosis includes a suppression of gut buddies such as *Lactobacilli* and *Bifidobacteria*, which produce dopamine and serotonin. As you know, these neurotransmitters help us feel good and play a role in the reward circuitry in the brain. You may recall that smoking also kills off *Bifidobacteria*, which makes smoking less and less rewarding and makes quitting more and more painful. The same exact thing happens with opioids, and this process prolongs and exacerbates addiction. When people who are addicted to opioids try to withdraw, the opioid microbiome resists the change, making it much more likely that they will ultimately relapse.[45]

It may sound like this is an impossible cycle to break, and it is indeed difficult, but there is plenty of hope. Probiotics have helped patients recovering from opioid addiction,[46] and in our program, we are going to heal your gut and foster a healthy, balanced inner terrain. This will make you less susceptible to addiction and/or better able to break these vicious cycles and fully and permanently recover. Just remember, these addictions are gut-driven and can be healed and controlled by focusing attention not on your brain's faults, but on your gut terrain's imbalance.

CHAPTER 7

MENTAL HEALTH AND THE GUT

Not long after the debate about evil bacteria versus the state of our inner terrain between Pasteur and Béchamp, a physician named António Maria de Bettencourt Rodrigues became the first doctor (that we know of) to make the connection between bacteria and mental health. Rodrigues was a Portuguese physician who was based at the Faculty of Medicine in Paris and had trained under the French psychologist Georges Dumas. At a mental health congress in Paris in 1889 (yes, 136 years ago!), Rodrigues presented the idea that depression could be caused by "autointoxication," which was the theory that the waste products of bacteria could back up in the body and cause disease.[1] He shared many examples of how he had successfully treated depressed patients with a combination of dietary changes and the elimination of toxins in the gut.

The idea of autointoxication went hand in hand with Pasteur's belief that all bacteria were harmful, and it became the prevailing theory of the day. Throughout the nineteenth century, autointoxication was widely believed to contribute to or cause many, if not all, diseases. This led to interventions such as colonic irrigation to clear out those bacterial waste products. This practice is still commonplace today![2]

Rodrigues may have been the first to publicly present on the idea that autointoxication could cause depression, but other physicians at around this time were also noting a potential connection between mental health and the gut. A scientist named François-André Chevalier-Lavaure noted that many of his psychiatric patients also suffered from digestive conditions, and that they saw improvements in their psychiatric symptoms after their issues with digestion were treated.[3,4]

Although the term "autointoxication" was incorrect and they got some of the details wrong, these scientists were definitely onto something. The mechanism behind the theory of autointoxication is shockingly pertinent to our discussion. However, the relatively new field of psychology grew in prominence over the course of the twentieth century, and the gut/mental health connection was mostly abandoned by the medical profession in favor of the idea that the mind and body were basically disconnected and should be treated separately. All these years later, we are finally waking up to the idea that many different types of mental health conditions are indeed directly related to the gut, in ways that continue to surprise and intrigue me.

Big surprise, the vast majority of these mental health conditions are impacted and exacerbated by—if not directly caused by—neuroinflammation. Numerous studies have noted the connections between leaky gut, a leaky BBB, increased cytokines, microglial activation, and mental illness.[5,6,7,8]

For one thing, when mice are given LPSs, those pieces of dead bacterial cell walls, it leads directly to anxiety-related behaviors.[9] Further, adults with major depressive disorder (MDD), schizophrenia, and bipolar disorder all have increased cytokine levels as well as higher-than-normal amounts of mitochondrial contents in their bloodstreams.[10,11] As I said earlier, those mitochondrial contents have been released from the cells during apoptosis. Because of their bacterial DNA, they are then recognized by the immune system and believed to be invasive bacteria. This triggers neuroinflammation.[12,13]

And while neuroinflammation is implicated in all the mental health issues we'll discuss, the details do vary. So, let's take a closer look.

THE DEPRESSED MICROBIOME

Rates of depression were already steadily increasing before the COVID-19 pandemic, which exacerbated this mental health crisis even more.[14] I don't mean to downplay the effects of the pandemic on people's mental health or the other factors that contribute to depression and other mental health issues. However, for the purposes of this book, I am choosing to focus on the connections between depression and the gut—of which there are many!

So, let's take a look at how the gut is contributing to depression. First and foremost, our gut bacteria can lead to depression via neuroinflammation stemming from dysbiosis and leaky gut. Dysbiosis can also lead to a change in neurotransmitter levels, which can lead to depression. In addition, a change in neurotransmitter levels can also cause inflammation, and inflammation can also cause worsening dysbiosis and leaky gut! There's that bidirectional relationship again.

Although the role of neurotransmitters in depression is widely known and researched, many people leave the microbiome out of this discussion. But as I discussed earlier, your gut buddies either produce your neurotransmitters directly or produce the precursors that your body needs to make those neurotransmitters. Simply put, with an altered terrain, neurotransmitter levels become dysregulated. This is a major factor in depression.

Indeed, patients suffering from major depressive disorder (MDD) have significantly altered microbiomes compared to healthy patients. Notably, patients with MDD have enriched pro-inflammatory bacteria and depleted anti-inflammatory, butyrate-producing bacteria.[15,16,17,18] Patients with MDD also have a higher chance of suffering from leaky gut and higher levels of circulating cytokines.[19]

Further, in 2022, a groundbreaking study looked at the microbiomes of over a thousand patients with depression. They found alterations in thirteen specific bacteria that were associated with depression. These specific bacteria are all known to be involved in the synthesis of glutamate, butyrate, serotonin, and GABA.[20] And in patients with severe mental illness, increases in dysbiosis, levels of zonulin (which causes leaky gut), LPSs, and inflammation are all correlated with the severity of disease.[21]

It also comes as no surprise that patients with other chronic inflammatory disorders also have higher-than-normal rates of depression. Inflammation doesn't discriminate. Once your immune system gets the message that the fortress of your gut lining has been breached, that inflammation will become widespread and eventually get to your brain.[22]

Of course, all of these alterations to the inner terrain cause dramatic differences in the messages being sent to the brain. One study looked at the metabolites in the blood of three groups of people: those suffering from depression, those who were in remission from depression, and a control group. The people suffering from depression had significantly higher levels of glutamate and alanine and significantly lower levels of myo-inositol, GABA, phenylalanine, creatine, methionine, oleic acid, and tryptophan compared to the other two groups.[23]

Balance is key when it comes to all of these metabolites. Either too much or too little can cause problems. I call this the Goldilocks effect. A balanced inner terrain should produce just the right amount of these substances. Clearly, in the case of patients suffering from depression, their microbiomes are not.

These low levels of tryptophan in patients suffering from depression are especially notable. Tryptophan is a precursor to serotonin and plays an additional role in the synthesis of serotonin. Specifically, tryptophan is also a precursor of 5-hydroxytryptamine (5-HT), an amino acid needed to activate serotonin receptors. Do those receptors sound familiar? That's because they are the ones that selective serotonin reuptake inhibitors (SSRIs), the predominant treatment for depression, work their magic on. Big shock—low 5-HT levels are a major cause of depression.[24]

But it's not a shock, actually. Those antidepressants actually work their magic not so much on the serotonin receptors themselves, but on (drumroll, please) your microbiome! SSRIs change the makeup of the gut and even have direct antibacterial effects.[25] One study compared the microbiomes of patients with MDD before and after they had been treated with an SSRI. Before treatment, their microbiomes were significantly different from those of healthy controls, with reduced diversity and richness. After treatment, their microbiomes had "normalized" and were much more similar to those of healthy patients.[26]

You may be wondering, *If SSRIs work by fixing the microbiome, then why not just go straight to the source and . . . fix the microbiome?* While pharmaceuticals are needed in some cases, this is exactly what we're going to do. Furthermore, think about this: If SSRIs really worked by increasing serotonin levels in the brain, they should kick in immediately, right? Yet patients and physicians know that the effects of these drugs usually take a month to kick in. Why? It takes that long to change your inner terrain.

TREATING DEPRESSION VIA THE GUT

Treating depression via the gut has been shown to be possible through several different mechanisms. These include fecal microbiota transplants, prebiotics and probiotics, hydrogen sulfide, and vitamin D. Even psychedelic treatments, which are now getting a lot of attention for their potential benefits, work by impacting—you guessed it—the gut.

Fecal Microbiota Transplants

A review of twenty-one different human studies found that fecal microbiota transplants from healthy patients given to those suffering from depression consistently led to a decrease in depression and anxiety symptoms. While it was not the goal, a recent review also found that transplanting fecal microbiota from patients suffering from depression to healthy patients led to depression and anxiety symptoms in those previously healthy patients.[27] Whoops!

Probiotics and Prebiotics

Probiotics can help regulate serotonin levels and reduce inflammation, leading to a reduction in symptoms of depression.[28] And prebiotics (which, you recall, are food for our gut buddies) also have antidepressant and antianxiety effects.[29] A recent review looked at human studies from 2015 to 2023 on probiotics and prebiotics being used to treat depression. The conclusion was that by attenuating inflammation and making serotonin more available, both prebiotics and probiotics significantly improved mood and reduced severity in patients suffering from depression.[30]

Hydrogen Sulfide

Hydrogen sulfide (H_2S) is another substance with antidepressant effects. In a fascinating study, mice were treated with LPSs. This led to an increase in neuroinflammation and the mice exhibiting symptoms of depression. Both the inflammation and the symptoms were reversed with the administration of H_2S![31]

So, what is H_2S, anyway? It's another signaling molecule made by your gut buddies, specifically when they ferment sulfur-containing compounds. I previously wrote about this in *Gut Check*, but it bears repeating. H_2S plays an important role in nociception, which is your nervous system's process of understanding noxious stimuli (heat, cold, mechanical force, or chemical stimulation).

When you experience pain, your gut buddies produce H_2S and send it to your brain to let them know that you're hurt. The H_2S then activates nociception neurons in the brain, which leads to the release of inflammatory cytokines and growth factors to heal the damage.[32] It makes sense, then, that disrupted nociception signals are associated with significant alterations in the microbiome. And when nociceptors are removed, the result is a defective tissue–protective reparative process.[33] This happens because neurons do not get the signal that you're in pain and need to heal.

This process of nociception provides the basis for "gut feelings" or "gut instincts." Our gut buddies produce H_2S, allowing us to interpret our own pain and discomfort. So, it absolutely makes sense that there is a direct link between H_2S and depression. However, this is another case where the Goldilocks effect applies. You don't want too much or too little H_2S. In fact, H_2S was historically considered a toxin because too much of it inhibits mitochondrial function.[34] But now we know that too little of it can lead to depression.[35]

Vitamin D

Increasing vitamin D levels is another potential treatment for depression, as depression and vitamin D deficiency often co-occur.[36] This also comes as no surprise, since vitamin D has a positive impact on the gut. It increases gut diversity and the relative abundance of butyrate-producing bacteria, as well as some other particularly helpful gut buddies like *Akkermansia* and *Bifidobacterium*.[37,38] The result of all these changes is less leaky gut and less inflammation.[39] This is precisely why vitamin D also helps prevent dementia[40]—something we'll talk about much more later on.

Psychedelics

Much like SSRIs, psychedelic drugs work with the gut to help alleviate depression in some downright fascinating ways. Ketamine, for example, can profoundly reduce inflammation by making changes to the microbiome.[41] When depressed mice were injected with either ketamine or saline daily for seven days, the ketamine group had a significant increase in friendly gut buddies and a reduction of pathogens.[42] Even better, after the first day's injection, the ketamine restored the animals' depressive-like alterations.[43]

My favorite part about this particular study is the fact that the mice's depressed state was created with none other than LPSs! So, the mice were given LPSs, which caused inflammation, which caused depression. Ketamine basically reverse engineered this process by healing the microbiome and the gut wall, which reduced inflammation, which helped reduce depression.[44] Pretty neat.

In addition, it's important to note that our gut buddies digest psychedelic substances for us, so they play a big role in the bioavailability of these drugs and their effects. In other words, the fact that we each have such unique inner terrains may explain why the effects of psychedelics tend to be so personalized, varying quite a bit between individuals, and subjective.

When you take psilocybin or "magic mushrooms," for instance, your gut buddies initially break them down for you and produce the metabolite psilocybin. Then with further digestion, this becomes psilocin, which is psychoactive and can cross the BBB.[45] Psilocybin and psilocin are tryptophan indole-based monoamine alkaloids. That's quite a mouthful, but the important thing to understand is that they work in your body much like tryptophan.[46]

As you recall, tryptophan produces indoles that promote neurogenesis.[47] Psilocybin and psilocin are structurally similar to these other indoles. They help alleviate depression by encouraging the

growth of connections between new neurons and by targeting serotonin receptors.[48,49]

It sure looks to me like patients with depression who don't have
the right gut buddies to make serotonin and indoles for them are
really benefiting from the way psychedelics can produce similar compounds. However, patients without the right gut buddies to break
down those psychedelics to begin with aren't able to produce psilocin
and experience the effects of these drugs.[50] That's a bit of a catch-22,
and the same is true of other types of psychedelics.

For instance, *Bifidobacterium*, which you recall is closely linked
to dopamine and addiction, modulates the metabolism of DMT, the
psychoactive compound in ayahuasca.[51] Another species of bacteria,
Enterococcus faecalis, produces a critical enzyme to degrade LSD.[52]
And there are specific bacterial strains that can metabolize mesca-
line, a psychoactive compound found in the peyote cactus.[53] Without
enough of these specific gut bugs, you can't fully benefit from these
drugs.[54]

Unfortunately, it appears that the very people who need these
drugs the most are the ones who are least able to benefit. In other
words, these patients have dysbiosis and leaky gut, which cause inflammation and depression and also cause them to be unable to metabolize psychedelics. This is just one reason I think it's a much better
idea to heal the microbiome before attempting to treat depression
with any of these substances.

CAN A DISORDERED MICROBIOME CONTRIBUTE TO AN EATING DISORDER?

It makes intuitive sense that suffering from an eating disorder would
cause changes to the microbiome, and this is true. But is it possible
that a disrupted inner terrain could contribute to or lead someone to
develop an eating disorder to begin with? I'm going to argue that the

answer is yes. First, let's look at how bacteria shape our dietary preferences in general—and vice versa.

To a great extent, we shape our microbiomes through the foods we eat. But why and how do we choose which foods and how much of them to eat? As you already know, our inner terrain is calling a lot of these shots. They send us messages telling us whether or not and how much to eat, as well as exactly which nutrients they need to grow and thrive. And we naively think that we just happen to be hungry and/or craving those foods.

I'm not just talking about one-off situations like a dash to the freezer for ice cream because of a hungry sugar-seeking microbiome. Rather, the makeups of our microbiomes have been shown to help dictate our hunger and our long-term dietary patterns. For example, the bacteria *Prevotella* thrive on carbohydrates, and people with a high abundance of this gut bug tend to eat high-carbohydrate diets. This, of course, allows *Prevotella* to grow and thrive and demand even more of those carbs. Meanwhile, *Bacteroides* thrive on protein and animal fat, and people with a predominance of these bacteria tend to be devout carnivores.[55]

Believe it or not, this can influence aggression. High-protein diets in dogs are strongly correlated to aggressive behavior, and low- to moderate-protein diets or the addition of tryptophan (remember, good gut buddies make this) promotes more social behavior.[56] A study on teenage girls in Iran in 2022 found that eating a Western diet was strongly associated with more aggressive behavior.[57]

Again, this is one reason it's so important to have a robust and diverse inner terrain—it prevents one species from overgrowing, demanding more of the food it wants, growing even more, and crowding out the other species while playing an outsize role in our dietary choices (not to mention other behaviors). With all sorts of different bacteria craving all sorts of different foods, a more diverse inner terrain leads to a more diverse diet and vice versa. It is really a beautiful system, if you think about it.

Of course, you may be wondering, *Is it the chicken or the egg?* In other words, doesn't someone have a predominance of a certain bacteria *because* they eat a lot of the foods that bacteria like, allowing them to thrive instead of the other way around? This is an excellent question, and the answer is both yes and no—chicken and egg, so to speak.

There's no arguing the fact that our diets do help shape our microbiomes, but the full picture is far more complex than that. For example, let's say that the bulk of your diet consists of carbs. You likely have a predominance of *Prevotella*. But where did it start? Were you born loving carbs and only carbs, and so the *Prevotella* in your gut multiplied while other species died out? Or did you start out with a predominance of *Prevotella* that sent messages to your brain, telling it to eat carbs, and you complied, which allowed them to multiply while other species died out? Given what we know about the infant microbiomes, I'm tempted to say it's the latter.

For instance, an interesting study looked at the effects of chocolate on the microbiome and mood. For three weeks, subjects ate 30 grams of chocolate per day, while a control group ate no chocolate (the horror!). After three weeks, the subjects in the chocolate group had significantly higher microbiome diversity than the control group. The chocolate-eating group also saw an improvement in their emotional states.[58]

Eating all that chocolate certainly made someone happy. But don't forget who produces the neurotransmitters that make us feel good. Our gut buddies, of course. They were happy with the chocolate they ate and produced feel-good chemicals as a reward.

Another study on chocolate had people follow identical diets and then placed them in two groups based on whether or not they tended to crave chocolate. The people in the chocolate-craving group had different metabolites in their urine compared to the people who were indifferent to chocolate (aka monsters).[59] Remember, they were eating the same things. So, did some of those metabolites travel up the vagus nerve and tell the brain to eat chocolate, creating the craving? I would say yes.

There is mounting evidence that messages being sent up the vagus nerve control our appetites far more than we've ever considered. When the vagus nerve is blocked, people tend to lose a drastic amount of weight.[60,61] And a study of girls suffering from the eating disorder anorexia nervosa showed that their vagus nerves were 30 percent more active (with messages from the gut to the brain) than girls who did not suffer from an eating disorder.[62]

In addition to all of the other signaling devices we've discussed, it turns out that our microbiomes can also literally change our taste receptors. In one experiment, germ-free mice were shown to have altered taste receptors for fat in their intestines compared to mice with normal microbiomes.[63] In another, germ-free mice preferred sweets and had more sweet taste receptors in their intestines compared to normal mice.[64]

Our microbiomes can also manipulate the hormones that control our sense of hunger and satiety. Germ-free mice have lower levels of leptin, the hormone that signals to your brain that you are full and should stop eating.[65] Even more intriguing, bacteria produce peptides that are molecularly nearly identical to leptin and other appetite-regulating hormones. These bacteria-produced peptides can mimic or block the effect of hormones by acting on their receptors.[66]

Further, I've been arguing that neuroinflammation is behind every mental health condition, and, indeed, this includes eating disorders. In fact, there is evidence that the neuroinflammation triggering anorexia comes from LPSs![67] We also know that patients with eating disorders, including anorexia, bulimia nervosa, and binge-eating disorders, tend to have altered neurotransmitter activity—specifically, an imbalance between the serotonin and dopamine pathways that are greatly controlled by the microbiome.[68,69,70,71] Patients suffering from anorexia are rewarded by their microbiomes for not eating, while those with binge-eating disorders and bulimia are rewarded for bingeing.

I treat many people with eating disorders. Let me share the recent example of my patient Sarah. Thanks to the recent advances in gut microbiome and intestinal permeability testing, this became a very clear case.

Sarah was an extremely bright and striving (common characteristics among my patients with eating disorders) high schooler with severe anorexia nervosa. After stays in many in-patient treatment centers around the country, she consistently relapsed shortly after returning home. Her parents brought her case to me when she was hospitalized and on a feeding tube for nourishment.

Sarah's blood work showed profoundly low levels of vitamin D and omega-3 fatty acids, as well as severe leaky gut and intestinal dysbiosis. In fact, she had a near-complete absence of SCFA-producing bacteria and therefore very low SCFAs and an overgrowth of pathogenic bacteria. I started Sarah on high-dose vitamin D$_3$, omega-3 fatty acids, appropriate probiotics, prebiotics including milk oligosaccharides and nano-encapsulated butyrate, and gut-wall-healing supplements. Within weeks, she was off the feeding tube, gaining weight, and compliant with her new eating regimen. She was surprised to find that for the first time in years, she actually enjoyed and looked forward to eating.

Of course, it wasn't a surprise to me at all. Béchamp and Bernard would merely nod and say, "Of course! Her terrain had changed; what else would you expect?"

If you or someone you love is suffering from an eating disorder or any mental health condition, it's important to remember it's not their fault—or yours. There is great freedom in accepting the fact that these tiny microbes are calling the shots—especially since we can manipulate them to change the ways that they are manipulating you.

CHAPTER 8

NEURO-INFLAMM-AGING

AND THE GUT

I hope you are beginning to absorb and accept the fact that the state of your inner terrain plays an enormous role in how you eat, think, behave, and feel. To a large extent, it also determines how your brain will age and whether or not you will suffer from neurodegeneration. It's been about a quarter of a century since the idea caught on that chronic inflammation is a major cause of aging, giving rise to the catchy and popular term "inflamm-aging."[1] At the time, however, there was no acknowledgment of where that inflammation was coming from, which of course we know now is the gut.

As you're about to see, chronic neuroinflammation stemming from dysbiosis and leaky gut is at the root of neurodegenerative diseases, including the two most common—Alzheimer's disease (AD) and Parkinson's disease (PD).[2] As with many of the other conditions we've explored, the number of people suffering from these and other neurodegenerative diseases has dramatically increased in recent decades, coinciding with the massive disruption of our inner terrain.[3]

Inflammation stemming from the gut drives neurodegeneration in a variety of ways. One is by making the blood-brain barrier (BBB) more penetrable.[4] As you read earlier, another is by leading the glial cells that protect neurons to cut off their communication with surrounding neurons. A third is by killing cells, including neurons, outright.[5]

In addition, both AD and PD also feature the accumulation and aggregation of misfolded proteins. To become biologically functional, the proteins in your body are "folded" from a chain-like structure into a three-dimensional structure. Think origami. When proteins are misfolded, which normally happens from time to time, they become dysfunctional.[6]

In a healthy cell, you have "quality control" systems to account for occasional misfolded proteins. If these cellular "quality control" system find a misfolded protein, they degrade and dispose of it, and all is well.[7] But when too many proteins are misfolded, these quality control systems become overwhelmed.

I like to imagine that classic *I Love Lucy* bit, where she is working in a chocolate factory and can't keep up with the number of chocolates racing toward her on the conveyor belt. While Lucy ate as many of the chocolates as she could and hilariously tried to stuff the rest in her hat and in her dress, misfolded proteins, like excess candies, have nowhere else to go. They start to build up in your cells, leading to organelle dysfunction, cell death, and eventual neurodegeneration.[8]

What causes so many proteins to be misfolded? you ask. Good question, and the answer is probably obvious by now—dysbiosis.[9] Further, when your inner terrain is disrupted, your gut is unable to produce essential signaling compounds like SCFAs, indole, brain-derived neurotropic factor (BDNF), and NO, which all tell your brain to produce new neurons via neurogenesis. In this state, as you age your neurons die and are not replaced—not a great recipe for longevity.

In addition to disease, neuroinflammation also causes the mental fatigue, or brain fog, that many of us have become so accustomed to that we consider it normal.[10] The truth is that you don't have to live with mental sluggishness, even as you age. When your inner terrain is able to function at its peak, so is your brain.

Studies indeed show that a healthy inner terrain is associated with longevity. For instance, centenarians and semi-supercentenarians have a higher relative abundance of our favorite gut buddy, *Akkermansia*, as well as additional helpful gut buddies like *Bifidobacterium* and *Lactobacillus*. And note that's not compared to the elderly. These centenarians and semi-supercentenarians have more of these gut buddies than healthy young people![11] Big shock—each of these specific bacteria helps maintain the gut lining.[12,13,14,15] As I said earlier, death begins in the gut. But never fear—so does the fountain of youth.[16]

THE ALZHEIMER'S DISEASE MICROBIOME

Alzheimer's disease was first written about and characterized in 1907 by two separate scientists who each described the presence of plaques and tangles in the brains of patients with AD. Indeed, AD is characterized by the accumulation of misfolded amyloid beta plaques between neurons that are tangled together by tau proteins. Interestingly, way back then, both of these scientists noted the similarities between AD and the bacterial infection neurosyphilis.[17] Yet, in the more than a hundred years since, we have failed to make the connection between bacteria and AD. Until now, that is!

We now know that amyloid beta is actually an antimicrobial peptide that is part of the brain's immune response.[18] It is produced in your brain and is meant to protect the brain from pathogens. When the immune system is constantly being activated because of dysbiosis and leaky gut, you end up with chronic inflammation and excessive amyloid beta production.[19] The result is much like Lucy working on that conveyor belt, unable to keep up. In the case of the brain, all that excess amyloid beta accumulates in neurons, where it becomes damaging and even lethal.[20]

In addition, when bad bugs in your gut overgrow due to dysbiosis, they can secrete their own bacterial amyloids that cause oxidative stress and induce the aggregation of proteins.[21,22] Bacteria produce amyloid proteins to help them bind to one another, forming biofilms.[23] This helps them resist being killed off by the immune system.[24] When bad guys, or any bacterial population that is overgrowing, stick together this way, it makes them more resilient, and therefore more dangerous.

Complicating this even further, although bacterial amyloids are not exactly the same as the amyloids produced in your brain, they do share a similar structure.[25] When your immune system is activated over and over by these bacterial amyloids, it becomes hyper-alert and begins attacking anything that resembles those bacterial amyloids— namely the innocent amyloids that are produced in the brain.[26] This is a case of what's called molecular mimicry, leading to an inflammatory autoimmune response, this time in your brain![27]

There is a great deal of evidence of AD's connection to an over-active immune system. For instance, patients with AD are more likely to carry latent viruses such as herpes simplex 1 (HSV-1) than those without AD.[28] Further, 90 percent of people with AD carry the respiratory pathogen *Chlamydia pneumoniae*, compared to 5 percent of healthy controls.[29] But don't worry, not all patients with HSV-1 (which is incredibly common) or who are exposed to *Chlamydia pneumoniae* will develop AD. So, what makes the difference? It's dysbiosis and leaky gut, of course.

There is a marked difference in the microbiomes of patients with AD compared to healthy people when it comes to their levels of microbial diversity and the relative abundance of specific species.[30,31,32] Notably, AD patients consistently have a reduced abundance of bacteria that produce butyrate and a higher-than-normal abundance of pro-inflammatory bacteria.[33] When mice are given fecal microbiota transplants from patients with AD, they develop amyloid beta plaques, neuroinflammation, and cognitive impairments. This, of course, does not happen when they are given fecal microbiota transplants from healthy donors.[34] And when mice are later given antibiotics to wipe out their "Alzheimer's microbiome," the amyloid beta plaques and cognitive impairments improve.[35] Whoa!

Dysbiosis also leads to an increase in a metabolite secreted by bad bugs called trimethylamine N-oxide (TMAO). TMAO contributes to the accumulation of amyloid proteins in the brain, increasing neuroinflammation and furthering the progression of AD.[36] It also induces mitochondrial dysfunction, oxidative stress, deterioration of neurons, and synaptic damage in the brain.[37]

But let's not forget about the one-two punch of dysbiosis and leaky gut, which is clearly at play in cases of AD. Patients with AD have higher levels of bacterial translocation and have three to six times more LPSs than healthy controls.[38,39,40] The immune system's response to all those LPSs flowing through a leaky gut drives chronic inflammation throughout the body,[41] ultimately leading to neurodegeneration[42] and AD.[43]

For further evidence that LPSs drive AD via neuroinflammation, consider this: Rats that were given a shot of LPSs developed amyloid beta and tau protein tangles within seven days.[44] Even the blood of patients with AD has a different microbiome profile compared to healthy controls.[45]

Because they are lacking butyrate-producing bacteria and have penetrable BBBs, bacteria and LPSs that have translocated from the gut can get into the brains of patients with AD, which have five to ten times more bacteria than healthy brains.[46] They also have higher-than-normal levels of LPSs in and around the neurons that have been affected by the disease.[47] In fact, the amyloid plaques that are a hallmark of AD often contain both bacteria and LPSs.[48,49]

FAT FOR BRAINS

At least half of the dry weight of your brain is made of fats, and the makeup of your brain's fats plays a big role in how well it functions and ages. Ideally, half of the fat in your brain should come from phospholipids, which contain both omega-3 and omega-6 fatty acids. The other half should be made of mostly glycolipids, which help maintain the stability of cell membranes, along with some cholesterol, triglycerides, and long-chain fatty acids.[50] Changes in lipid metabolism are associated with AD.[51] I'll give you one guess who plays a big role in fat metabolism and the makeup of your fatty brain—your gut buddies, of course.

For one thing, patients with AD tend to be lacking an important type of lipid in the brain called plasmalogens, and low levels of these lipids are associated with chronic inflammation.[52] Amazingly, your gut buddies can convert inulin, a sugar in chicory, into plasmalogens.[53] This is just one small way that having the right gut buddies (and the right diet) can help protect you from AD.

Further, you read earlier that alpha-linolenic acid (ALA), an essential fatty acid found in only certain plant oils like perilla and flaxseeds, helps protect you from the endotoxemia associated with a high-fat diet by feeding your gut buddies while keeping bad bugs from overgrowing.[54] Sure enough, ALA has been shown to help degrade tau proteins, reduce neuroinflammation, repress neuron loss, and enhance memory and cognitive function.[55,56,57,58] Not bad!

Even the biggest genetic risk factor for AD—the apolipoprotein E4 (APOE 4) genotype—all comes down to your inner terrain. Most people carry two copies of the APOE 3 gene, but there is a mutation that causes people to carry APOE 4. About one in four people carry one copy of APOE 4, which doubles the risk of AD. Far fewer people carry two copies of APOE 4, which raises the risk of AD more than twelvefold.[59]

The APOE gene helps metabolize fats, and people with APOE 4 don't produce enough apolipoproteins, which are proteins that transport lipids. The result is that cholesterol can't move in and out of cells properly and builds up inside of cells. In addition, docosahexaenoic acid (DHA) can't move into their brains.[60] This is a big deal, as DHA makes up half of the membrane content in our neurons.[61]

As I wrote in my last book, *Gut Check*, your genetics also play a role in determining the makeup of your inner terrain.[62,63] Your diet and lifestyle play a bigger role, but when you're dealing with a risk factor like APOE 4, it's important to take everything into account. In particular, the APOE 4 genotype is associated with a specific micro-biome profile that includes significant reductions in amino acids and SCFAs.[64] This means that even in the case of a genetic mutation, your fate is in the hands of your inner terrain.

Here's an example of this: A study from 2019 showed that Nigerians have higher-than-average rates of the APOE 4 mutation. Yet, compared to African Americans living in Indianapolis with the same mutation, these Nigerians suffer significantly less dementia and Alzheimer's disease.[65] Why? Well, the traditional Nigerian diet features a huge number of complex starches and tubers, while the Indianapolis diet (like most Western diets) is nearly devoid of these gut-buddy-friendly foods and is instead full of simple sugars and saturated fats. Having a balanced inner terrain protected the Nigerians, even with the genetic mutation.

As with many of the seemingly scary topics I've already covered, I believe this connection between the microbiome and Alzheimer's offers a lot of hope. As I said, you have a tremendous amount of control over your inner terrain, even if your genetics predispose you to having a particular microbiome profile. For example, manipulating the microbiomes of mice led to significant decreases in inflammation, tau pathology, and resulting damage from AD.[66] And people with AD, whether they have the genetic mutation or not, benefit from taking fish oil (which contains DHA) and apolipoproteins.[67]

You can truly take back control of your mind by taking control of your inner terrain, even when it seems the deck is stacked against you.

THE PARKINSON'S DISEASE MICROBIOME

Symptoms of Parkinson's disease (PD) are caused by the degeneration of nerve cells in the part of the brain called the substantia nigra, which controls movement. This is why patients with PD often suffer from tremors and other movement issues. The substantia nigra is also a critical brain region for the production of dopamine. When patients with PD lose nerve cells in the substantia nigra, they lose the ability to produce dopamine.

Perhaps the dopamine connection to PD should have been our first clue that, while symptoms of PD may come from the death of these neurons, PD itself really stems from the gut. We already know that our gut buddies play a huge role in our dopamine production. Or perhaps this clue should have been first—gastrointestinal dysfunction is a common early symptom of PD.[68] In fact, GI symptoms often appear before patients develop any of the motor symptoms that we associate with PD.[69] This makes sense, since dysbiosis and leaky gut come first, driving neuroinflammation and the death of cells in the substantia nigra, leading to motor symptoms.

Well, whichever clue you prefer, it's now clear that PD is yet another symptom of dysbiosis and leaky gut leading to systemic and neuro-inflammation.[70] A human study of patients with PD showed that the disease is associated with significant dysbiosis, including an altered abundance of over 30 percent of bacterial species, genes, and pathways.[71]

Like patients with AD, those with PD also tend to have significant reductions in butyrate-producing bacteria and other important anti-inflammatory gut buddies.[72,73] They also have reduced expression of tight junction proteins, causing or worsening their leaky guts, and higher-than-normal concentrations of LPSs and inflammatory markers.[74,75] Another well-designed human study showed that higher plasma levels of LPSs are associated with an increased risk of PD and that endotoxemia plays a role in the pathogenesis of the disease.[76]

Similar to AD, a hallmark of PD is the buildup of misfolded alpha-synuclein (a-Syn) proteins that accumulate in between neurons. These clusters of misfolded proteins are called Lewy bodies.[77] In addition to their brains, patients with PD also have Lewy bodies in between the neurons in their guts![78]

It certainly appears that PD begins in the gut and spreads to the brain via the vagus nerve.[79] Studies on patients who have received a vagotomy, which involves cutting one or more branches of the vagus nerve, show that a truncal vagotomy, which fully transects the nerve, reduces PD risk by 40 percent.[80] Indeed, without a vagus nerve, there is no "telephone wire" along which those Lewy bodies can travel from the gut to the brain.[81,82] The Lewy bodies can then breach the BBB because of the "Parkinson's microbiome," which includes a lack of butyrate-producing bacteria and hence a leaky brain.

Intriguingly, it has recently been discovered that patients with PD have significantly lower levels of niacin, a B vitamin, than healthy control subjects.[83] This seems to be a result of the disease itself that is worsened by PD medication. A side effect of the most common medication prescribed to treat the motor-related symptoms of PD is that it reduces the conversion of tryptophan to niacin.[84]

Did an alarm bell go off just now? I sure hope so. If you've been paying attention, you're likely already picking up on the fact that this medication is acting on the microbiome, as your gut buddies produce tryptophan, the precursor to niacin, as a postbiotic signal to the brain. Indeed, niacin supplements have been shown to shift the microbiome from a pro-inflammatory to an anti-inflammatory state.[85] Perhaps this is why probiotics taken together with prescription medications strengthen the medication's ability to manage PD symptoms. The probiotics heal leaky gut, restore the inner terrain, and offset the medication's side effects.[86]

* * *

At this point, I hope I have removed any doubt about who is really in control of your brain. I traveled down many, many rabbit holes while doing the research for this book, and every single one of them led me to the exact same place—that precious inner terrain. The bad news is that your once beautiful, rich terrain is now more likely to resemble a desert wasteland, or maybe worse: a smelly, overgrown swamp. The good news is that you can restore it back to its former glory, and your brain along with it. Now, that is exactly what we are going to do.

CHAPTER 9

THE DOS AND DON'TS OF THE

GUT-BRAIN PARADOX PROGRAM(S)

By now, I hope you have accepted the fact that your microbiome is calling the shots when it comes to your appetite, behavior, cognition, brain aging, and so on. The good news is that you get to call the shots about what you do and don't eat. This can drastically change the makeup of your inner terrain and therefore the messages that are being sent from your gut to your brain. So, let's take a look at how to best nourish the right gut buddies to ensure they are telling you to behave in ways that are good for both them and you.

This time around, I am going to offer two distinct programs that you can choose from based on your personal needs and goals. Most of you will choose the regular Gut-Brain Paradox Program, which will provide many options of delicious foods for you and your gut buddies to enjoy. However, for those suffering from addiction or significant mental health issues, I am also including a unique program that I call the Chicken and the Sea. As the name implies, this program relies primarily on chicken and seafood for reasons that will become clear. Of course, if you want to jump-start your progress, you can choose to start with the Chicken and the Sea Program and then switch to the regular Gut-Brain Paradox Program.

For both programs, however, our goals are the same. So, let's start there.

THE GUT-BRAIN PARADOX GOALS

First and foremost, our main goal is to reduce and prevent neuroinflammation. Period. This is how we will improve any brain issue you are currently dealing with and/or keep any new ones from occurring. How will we do this? First, by balancing the environment of your inner terrain. As Béchamp noted so many years ago, this will keep bad bugs from overgrowing and sending the wrong messages to your brain and will allow your helpful gut buddies to thrive. These gut buddies will then send messages telling your brain to be happy, eat healthy foods, and think sharply, all while telling your immune system to calm down. With increased stability of your terrain and interdependence among species of bacteria, the intricate systems in your gut will be able to function harmoniously, as they were designed to do.

Second, the Gut-Brain Paradox Program will help repair and protect the gut wall. The turnarounds I have seen in my patients just from restoring their gut walls are nothing short of remarkable. Repairing the gut wall reduces the amount of LPSs, pathogenic bacteria, and undigested foods entering the bloodstream. This, of course, reduces immune system activation and is key to reducing and preventing neuroinflammation. And don't forget that the same tactics that help repair the gut wall also help restore the BBB. This is a one-two punch of protection for your brain.

Indeed, a human study using a similar dietary plan to the Gut-Brain Paradox Program resulted in modulation of the microbiome, improved SCFA production, and significant positive changes to the endocrine system. This led subjects to lose an additional 116 calories a day through their stool![1] Talk about a caloric bypass.

Finally, this program will help keep your mitochondria healthy. This will also help protect the gut wall and the BBB. The healthier the mitochondria and therefore the cells along these barriers, the stronger and more resilient they will be.

How will we keep those mitochondria healthy? I'm glad you asked. While I covered this in several of my previous books, I want to take a brief detour here to explain an important topic: mitochondrial uncoupling. Feel free to skip ahead if this is familiar territory, or keep reading for a refresher. If you are new to my books, this is important information to have.

As you may know, mitochondria convert food and oxygen into energy via a process called cellular respiration, which is a bit like an assembly line. I'm simplifying this somewhat, but essentially, during this process, protons and electrons enter mitochondria, go through a series of chemical reactions, and couple up with oxygen molecules. When protons couple with oxygen, it creates ATP (energy) and leaves behind carbon dioxide as a by-product. When electrons couple with oxygen, it creates reactive oxygen species (ROSs), which damage your cells. Too many ROSs can trigger apoptosis, which you know can cause neuroinflammation.

To prevent too many ROSs from building up in the mitochondria, your mitochondria can open up emergency exits that are controlled by switches called uncoupling proteins. This allows protons to escape without producing ATP. Even better, the cell can create new mitochondria via mitogenesis, where those extra protons can then go to couple with oxygen and make energy.

This process cuts down on the amount of energy (and therefore ROSs) that each mitochondrion makes but increases the amount of energy being made overall in the cell. The reduction in ROSs helps keep the mitochondria and the cell healthy. Even better, your cells get the raw materials for building those new mitochondria by opening up your fat stores!

Mitochondrial uncoupling is the best way to protect the health of your mitochondria and cells, all while burning more fat. So, how do you make it happen? Specific compounds activate those uncoupling proteins that reside in the inner mitochondrial membrane and manage this process. You will be consuming plenty of these compounds on this program, which will protect the health of your cells, including the ones along your gut wall and BBB. Again, this will help dramatically reduce neuroinflammation and protect the health of your brain.

GUT-BRAIN PARADOX DOS AND DON'TS

To keep things simple, I am including a list of dos and don'ts to help you make good choices as you move forward with this new way of eating.

Do: Eat Your LPSs

Let's start with a new rule that I can barely believe I'm including. But that's right—I want you to start consuming LPSs to teach your immune system to recognize (and no longer be afraid of) bacteria. There are many ways for you to get a healthy dose of LPSs. You can do what I do and buy your root veggies at the farmers' market and avoid washing them before eating. Yes, I'm serious. Focus on the veggies that get pulled out of the ground like parsnips, radishes, turnips, and carrots. Okay, okay, I do rub off a lot of the dirt, but only because my wife makes me. (Just kidding, Penny.)

It's also a good idea to use plenty of spices in your cooking, not just for their flavoring purposes and polyphenols (more on that in a moment), but also as a delivery device for LPSs. Using lots of fresh unwashed herbs from the farmers' market is also a great way to get some LPSs. You can even try growing your own!

Do: Eat Fermented Foods

A diet high in fermented foods dramatically increases microbiota diversity while decreasing inflammatory markers.[2] Fermented foods have been shown to improve the integrity of the gut wall, leading to a reduction in inflammation and even anxiety.[3] For all these reasons, fermented foods are some of the most important ones you'll be eating on this program. In fact, they're even more important than those containing dietary fiber, which most people assume are the key to a healthy microbiome.

What most people misunderstand about dietary fiber is that in order to benefit from it, you need your gut buddies to digest it for you. And before your gut buddies can digest fiber for you, they need it to be "predigested" via fermentation. Perhaps even more important, through fermentation, yeasts and bacteria produce "intermediary SCFAs," which include formate, succinate, and lactate. These intermediary SCFAs are not for you. They are food for your bacteria, particularly those important butyrate-producing bacteria.[4] Fermented foods contain these intermediary SCFAs, and when you eat them, it allows your gut buddies to produce butyrate.

Another reason that fermented foods are so important is that they contain both living and dead bacteria, as well as the postbiotics that were produced by those bacteria when they were all alive. These three components each send distinct and important messages to your brain. One of the most important types of postbiotics in fermented foods is called polyamines. These compounds help modulate the immune response, protect the gut wall, and act as mitochondrial uncouplers.[5] Polyamines also increase the activity of intestinal alkaline phosphatase (IAP), a compound that breaks down LPSs so they can no longer cause you harm.

The best choices for fermented foods include plain sheep's and goat's milk yogurts, as well as low-sugar kombuchas; raw cheeses; unpasteurized (raw) cheeses from Italy, France, and Switzerland; apple cider vinegar and any other vinegars, pasteurized or not; sauerkraut; and kimchi. If you choose to eat meat, it's a good idea to make sure it is also cured, aka fermented. Studies show that artisanal fermented meats such as pancetta and prosciutto can act as powerful probiotics![6]

Do: Eat Polyphenols

Like fiber, polyphenols have been misunderstood for ages—and like fiber, we need our gut buddies to predigest or process them for us! Certain gut buddies love to eat polyphenols, and through their digestion, bacteria turn polyphenols into more absorbable and bioactive compounds.[7] This means that you can't fully benefit from polyphenols if you don't have a balanced inner terrain to digest them for you and make them active.

In addition, polyphenols and your microbiome have a bidirectional relationship. Your gut buddies process polyphenols for you, and polyphenols regulate the balance of buddies in your gut.[8] So, you should make sure to eat foods containing polyphenols while working to improve the balance of your inner terrain. You will benefit more and more as your gut buddies become better able to thrive and process those polyphenols.

Once they are processed by your gut buddies, polyphenols help the brain in several ways. First, they activate mitochondrial uncoupling and therefore protect your mitochondria.[9] This can help prevent and reverse leaky gut. The healthier the mitochondria in your gut wall, the stronger that wall will be.[10]

Certain polyphenols also send a signal to SIRT1, an enzyme in the cell nucleus that repairs and protects DNA from damage and is deeply involved in brain function. When polyphenols stimulate SIRT1, it enhances brain function and longevity.[11] With all these brain benefits, polyphenols have been shown as an effective treatment for ADHD, either alone or in combination with pharmaceuticals.[12]

One type of polyphenol that is particularly helpful for the brain is resveratrol (RSV), which can protect against Alzheimer's disease by inhibiting memory loss and creating neural malleability in the hippocampus.[13] RSV and other polyphenols can also activate AMP-protein kinase (AMPK), an enzyme that helps promote neurogenesis (the birth of new neurons) and mitochondrial biogenesis[14] (the birth of new mitochondria).

Another powerful polyphenol is called ellagitannins, which are present primarily in raspberries, pomegranate seeds, and walnuts. Your gut buddies can use these to produce a metabolite called Urolithin A. Urolithin A uncouples mitochondria and triggers mitophagy (the recycling of old, damaged mitochondria) and mitogenesis (the creation of new mitochondria).[15]

Unfortunately, only about 20 percent of the general population has the right mix of bacteria in their guts to produce Urolithin A from polyphenols.[16] This is why our primary goal is to restore balance to your inner terrain. Only then will you be able to fully benefit from ellagitannins and other beneficial polyphenols.

The good news is that polyphenols are present in some of the most universally adored foods, including wine, coffee, and chocolate. Perhaps this is why coffee helps improve memory![17] Just make sure you don't undermine the benefits with low-quality chocolate or by adding cream to your coffee. Choose dark chocolate and black coffee. If you feel that you need to add creamer, choose one of the many keto MCT creamers on the market.

You can also get polyphenols while consuming your daily dose of LPSs from herbs and spices. So, if you don't want to consume caffeine, try decaf or have an herbal tea made with polyphenols like mint. Or choose other polyphenol-rich plant foods like kiwi, pomegranate, red berries, or kale. Just remember to take the sugar levels of these foods into account before consuming them.

To make sure you benefit as much as possible from the power of polyphenols, however, it's a good idea to leverage the powers of fermentation and polyphenols together by eating fermented polyphenols.[18] This allows you to consume those polyphenols after they've already been processed and are more bioavailable.[19]

There are many ways to work fermented polyphenols into your

diet. Balsamic vinegar is a great source. I like to start my day by adding a dash of balsamic vinegar to sparkling water to make a refreshing sparkler. You can also add balsamic vinegar to your green smoothie, pour it on top of goat's or sheep's milk yogurt or kefir, or add it to your salad dressing.

Here's a list of common foods sorted by their polyphenol content, ranked from highest to lowest.[20] You might notice that a lot of these polyphenol-containing foods are spices that have LPSs, as well. A win-win. Try to find as many ways as possible to add fermented versions of these foods into your diet.

- Cloves
- Peppermint, dried
- Star anise
- Cocoa powder
- Mexican oregano
- Celery seeds
- Black chokeberries
- Dark chocolate
- Flaxseed meal
- Black elderberries
- Chestnuts
- Common sage, dried
- Rosemary, dried
- Spearmint, dried
- Common thyme, dried
- Lowbush blueberries
- Black currants
- Capers
- Black olives
- Highbush blueberries
- Hazelnuts

- Pecans
- Plums
- Green olives
- Sweet basil, dried
- Curry powder
- Sweet cherries
- Artichokes
- Blackberries
- Strawberries
- Red chicory
- Red raspberries
- Coffee, filtered
- Ginger, dried
- Prunes
- Almonds
- Black grapes
- Red onions
- Green chicory
- Common thyme, fresh
- Refined maize flour (masa made from hominy)
- Tempeh
- Apples
- Spinach
- Shallots
- Lemon verbena, dried
- Black tea
- Red wine
- Green tea
- Soy yogurt
- Yellow onions
- Pomegranate juice (100% juice)
- Extra-virgin olive oil

- Black beans (pressure-cooked or fermented)
- Peaches
- Blood orange juice (100% juice)
- Cumin
- Grapefruit juice (100% juice)
- White beans (pressure-cooked or fermented)
- Chinese cinnamon
- Blond orange juice (100% juice)
- Broccoli
- Red currants
- Pure lemon juice
- Apricots
- Caraway
- Asparagus
- Walnuts
- Potatoes (pressure-cooked)
- Ceylon cinnamon
- Parsley, dried
- Nectarines
- Curly endive
- Marjoram, dried
- Red lettuce
- Quinces
- Endive (escarole)
- Pumelo juice (100% juice)
- Organic rapeseed oil (canola oil)
- Pears
- Soybean sprouts
- Green grapes
- Carrots
- Vinegar
- White wine
- Rosé wine

Do: Eat the Right Fats

You read earlier about how important the right fats are for a healthy brain. In particular, phospholipids help protect the health of mitochondria. Mitochondria have double membranes composed primarily of phospholipids. These fats house uncoupling proteins. Simply put, phospholipids are essential for mitochondrial function and uncoupling. In fact, mitochondrial phospholipids regulate apoptosis. Yet, your mitochondria cannot produce these fats themselves.[21]

The best source to help generate phospholipids is the short-chain omega-3 fat alpha-linolenic acid (ALA), which is found in perilla oil, flaxseed oil, Ahiflower oil, and organic canola (rapeseed) oil.[22] ALA has profound anti-inflammatory properties,[23] and simply adding ALA to your diet can have a tremendous impact on your brain health.[24] Mice that were put on an ALA-enriched diet saw huge improvements to the makeups of their microbiomes.[25] ALA can help heal your gut and even your metabolism from the impact of a Western diet, while improving endotoxemia.[26,27] Perilla oil has also been shown to improve microbiome and gut wall function in athletes![28]

Preformed phospholipids are also plentiful in shellfish like mussels, scallops, clams, oysters, shrimp, crab, squid, and lobster. Omega-3 egg yolks have generous amounts of short-chain omega-3s, as well as being a rich source of arachidonic acid, a type of long-chain omega-6 fatty acid that is essential to your brain.[29] Olive oil is also a great source of the fats and polyphenols you need to keep your microbiome and brain in balance.

Do: Eat Your Veggies

Some of your most important gut buddies need veggies in order to grow and thrive, so they are a must to balance your inner terrain. If you don't love these foods now, the good news is that once you have more of these gut buddies, they will start sending messages to your brain telling you to eat more of them. Suddenly, you'll start craving salads instead of carbs—I promise!

Start by focusing on cruciferous vegetables like broccoli, cauliflower, and other sulfur-containing veggies, including onions, garlic, leeks, chives, shallots, and scallions. But all green, leafy vegetables are good choices, too. Better yet, if cruciferous veggies aren't your thing, look for the chicory family of leafy greens, which includes radicchio, Belgian endive, chicory, and frisée (also known as curly endive).

Do: Get Your Fiber

While I mentioned that dietary fiber has long been misunderstood, it is still important. Eating foods that are rich in soluble (and some insoluble) fibers, including tubers, rutabagas, parsnips, radishes, root vegetables, radicchio, endive, okra, artichokes, pressure-cooked beans and legumes, basil seeds, flaxseeds, and psyllium seeds, supports the health and reproduction of healthy bacteria in your microbiome. When your gut buddies get the sustenance they want, they'll send messages to your brain saying that their needs are being met. As a result, you will feel less hungry and begin to crave more of the foods, like those veggies I just mentioned, that are healthy for them and therefore healthy for you.

One of the very best prebiotics is inulin, a type of dietary fiber found in foods like chicory, asparagus, onions, leeks, and artichokes. As I mentioned earlier, one of the many benefits of inulin is that your gut buddies use it to produce plasmalogens that protect your brain.

Another important type of fiber is prebiotic fiber, such as psyllium husk powder, which helps promote gut buddy diversity. One of my favorite sources of prebiotic fiber is soaked basil seeds. As many of you know, chia seeds are loaded with lectins. But basil seeds give all the benefits of forming a prebiotic gel without lectins. Start with a teaspoon a day, either mixed in water or not, and work up to a daily tablespoon or even two.

Finally, don't forget your animal fiber. That's right—this is the food for gut buddies that is lurking in some animal products. Unfortunately, it's the stuff that most of us eating a Western diet discard without eating, such as the ligaments, tendons, bones, and cartilage in meats and fish. This stuff is important, however, because it resists digestion in the small intestine and then gets gobbled up by your gut buddies in the large intestine, allowing them to grow and thrive.[30]

One of the cornerstones of "ancestral diets," followed primarily in countries with historically high rates of poverty *and* longevity, is the concept of eating "nose to tail." This means that they regularly consume every part of the animal to avoid waste. It turns out that this also protects their brains. Moreover, in my travels studying these cultures, I have been impressed by the wealth of fermented animal products, like sausages, hams, and cheeses. Now you and I understand the mechanism of action and can make it a part of your brain-stimulating diet.

Do: Eat Melatonin

Melatonin is a hormone that has also been widely misunderstood. It is mostly known to induce sleep, and it can indeed make you sleepy. But more important, it is an antioxidant that helps protect the health of your mitochondria.

Melatonin is made in the brain from the amino acid tryptophan (which you recall is also a precursor to many important neurotransmitters). However, melatonin is also present in lots of different plant compounds, including leaves, stems, roots, fruits, and seeds. You can get your melatonin from many of the same foods that are also great sources of polyphenols, such as red wine, olive oil, mushrooms, nuts, and spices.

The following foods contain high levels of melatonin (listed from highest to lowest melatonin content). But, to be clear, please pressure-cook, cool, and reheat the types of rice below.

- Pistachios
- Mushrooms
- Black pepper
- Red rice
- Black rice
- Mustard seeds
- Olive oil
- Brewed coffee
- Red wine
- Cranberries
- Almonds
- Basmati rice
- Purslane (the weed the Ikarians eat)
- Tart cherries
- Strawberries
- Flaxseeds

Do: Take Your Vitamin D

Vitamin D is closely linked to longevity,[31] likely because of all the ways it helps protect and heal the gut. You need adequate levels of vitamin D for stem cells to proliferate and replace dead cells along your gut wall. This is essential for avoiding leaky gut. Higher vitamin D levels also lead to increased microbial diversity along with an increase in butyrate-producing bacteria[32] and other important gut buddies like *Akkermansia* and *Bifidobacterium*.[33] As a result of all this gut protection, high levels of vitamin D lead to reduced autoimmunity[34] and dementia.[35]

One of the leading vitamin D research groups at the University of California San Diego believes that the average person should take 9,600 IUs of vitamin D_3 daily to have safe and therapeutic levels. They have not seen, and nor have I, vitamin D toxicity up to 40,000 IUs daily.[36] Many of my patients with leaky gut initially require doses of 20,000 IUs daily. I aim for levels of 100 to 150 ng/ml on blood tests.

Of course, the most obvious source of vitamin D is the good old sun, which I urge you not to shun completely! And don't get me started on the dangers of most commercial sunscreens. Sunlight is a potent source of energy, but in order to convert it into ATP, we need chlorophyll. When we eat lots of greens and get healthy sun exposure, we get vitamin D, and our mitochondria can literally produce more energy![37]

Don't: Eat Foods That You Are Sensitive To

Because reducing neuroinflammation is of utmost importance, on this program you are going to stay away from any foods that you may be sensitive to. When you are sensitive to a certain food, it leads to an immune system reaction, which causes neuroinflammation. The good news is that once your leaky gut heals and your immune system stands down, you can slowly reintroduce these foods without causing neuroinflammation.

Ordinarily, the food you eat is digested by your digestive enzymes and your gut microbiome into individual molecules of protein (amino acids), sugar molecules (glucose or fructose) from carbohydrates, and fatty acids. With the exceptions of fats, these molecules are passed through each cell lining your gut and put into the large portal vein heading into your liver. If you have leaky gut, however, undigested pieces of food can pass through the wall and are seen as "foreign" by your immune system lining your gut. Your immune system then produces an antibody against that food so it can recognize it next time and attack. These antibodies are IgG and IgA, not IgE, which is the antibody that's formed in the case of a true allergy, not a sensitivity.

So, how do you know what foods you are sensitive to? In my clinics, we test for leaky gut and food sensitivities. After doing these tests on thousands of my patients, I've found that nearly 100 percent of all my patients have markers for leaky gut; 98 percent of tested patients have strong IgG antibodies to WGA (wheat germ agglutinin), gluten, and non-gluten components of wheat; and 70 percent react to proteins in corn, even those who haven't eaten these foods in years.

When you have leaky gut, lectins, bacteria, and undigested food are all crossing the gut wall into your body 24/7. The immune system believes that you are under constant attack and keeps all antibodies against foreign substances active and primed and ready. However, when my patients seal their leaky guts, all those antibodies to WGA, gluten, corn, and wheat disappear. I trust that the same thing will happen for you.

On this program, you're going to avoid wheat and corn, plus the following foods that my patients are most often sensitive to:

- Almonds and almond flour
- White mushrooms
- Ginger
- Pineapple
- Peaches
- White onions
- Lemons
- Bananas
- Nutmeg
- Cinnamon
- Commercial poultry
- Vanilla bean

Don't: Eat Unfermented Dairy Products

Speaking of food sensitivities, many people are also sensitive to cow's milk and all dairy products made from cow's milk. If this is you, consuming dairy products will activate the immune system and cause neuroinflammation. For instance, one study showed that cream intake led directly to an increase in plasma LPS concentrations, while intake of orange juice or water did not.[38]

The good news is that fermenting dairy products degrades the allergens, including casein.[39] Fermented dairy products are therefore a much safer bet. Plus, these products offer all the benefits of other fermented foods. Yogurts, for example, can inhibit inflammation, reshape the microbiome, improve glucose metabolism,[40] and improve endotoxemia.[41]

However, most American and northern European fermented cow's milk dairy products are still not a great idea. If you've read my other books, you already know that most cow's milk products in the United States come from a breed of cow that produces milk with a highly inflammatory protein called A1 beta-casein. Even fermented yogurts made from A1 cows are inflammatory. Plus, these conventional yogurts are usually loaded with sugar, and even if they are plain and unsweetened, they still have the wrong casein molecule.

If you can find sugar-free yogurts made with A2 milk, go for it. But in general, it's a much better choice to go for sheep's and/or goat's milk yogurts and cheeses.[42] As an added benefit, sheep's milk has over four times the amount of vitamin D as cow's milk![43] You can also look for water-based kefirs or coconut yogurt.

Don't: Eat Processed Foods

Foods that are derived from agricultural products and have been industrialized with additives not normally found in foods are considered processed. These foods have essentially been "predigested" and provide our gut buddies with no nourishment whatsoever. Eating these foods can kill off our helpful gut buddies and destroy our microbiome and our metabolism.[44]

Ironically, these "foods" are harmful because of what they both do and do not include. They do not include anything for our gut buddies to ferment. This literally starves them to death. And what they do include often kills those gut buddies outright—or, at least, those that haven't died of starvation yet.

Processed foods are typically full of polyunsaturated omega-6 fats from soybean and corn oils, which destroy your gut buddies' ability to make hydrogen sulfide, a gasotransmitter that helps alleviate inflammation in the gut.[45] Further, the short-chain omega-6 fat linoleic acid is the most prevalent fat in most industrial seed oils, like soy, corn, cottonseed, sunflower, safflower, and grape-seed oil. When heated, these are transformed into aldehydes, which are toxic to your mitochondria.

Processed and fried foods are also hidden sources of trans fats, which have been banned but somehow still sneak into our food supply. This type of fat, created through industrial processing, clogs and damages the inner membranes of your mitochondria.

Finally, processed foods are chock-full of chemicals, like food colorings, artificial sweeteners, and high-fructose corn syrup, which all damage our microbiomes. Just a single packet of an artificial sweetener like sucralose (Splenda) can kill off half of your gut buddies, and titanium dioxide, a common additive, has been shown to alter the composition of your microbiome and cause inflammation.[46]

For the first time, on this program, I am recommending that you avoid the one type of processed food that I used to allow, which is nutritional bars and protein powders. These foods are harmful to the brain in particular because they contain monosodium glutamate (MSG), used as a flavor enhancer. High intake of MSG is associated with chronic pain, likely because it drives neuroinflammation.[47] Remember, the bad bugs use pain to control your behavior. So, let's avoid any and everything that might contribute to this problem. Sadly, MSG does not have to be displayed on the ingredients list, so buyer beware.

Don't: Consume Much Fructose

As you read earlier, fructose, which is half of the sugar in table sugar and the main sugar in fruit, is bad news for your microbiome. It is a mitochondrial toxin, reduces microbiome diversity, feeds bad bugs, and contributes to leaky gut. Fructose also hampers gasotransmitter production,[48] literally changing the messages being sent from your gut to the brain!

Perhaps this is why sugar intake is associated with cognitive dysfunction. In the Framingham Heart Study, intake of soft drinks and fruit juices was associated with dose-dependent reductions in brain volume and poor episodic memory.[49] Sugar intake is also associated with poor cognition in older people.[50] And don't forget the fact that fructose and other foods, like juices and soft drinks, can impact you (or your children) before birth. For instance, maternal consumption of soft drinks during pregnancy and consumption of soft drinks during early childhood are associated with cognitive dysfunction in children.[51]

Meanwhile, all-natural fruit may be a good source of polyphenols and other nutrients, but most fruit is now bred to be sweeter than normal, with higher levels of fructose. For this reason, choose only low-sugar fruits when in season, such as pomegranate and passion fruit seeds, kiwis (with the skin on), and grapefruit.

Don't: Eat Lectins

As you well know by now, lectins cause leaky gut and can damage the BBB and other internal organs when they cross the gut lining. One particularly dangerous type of lectin is called wheat germ agglutinin (WGA). WGA is part of the wheat germ, which is normally removed during milling but is present in whole grains.

WGA is so small that it can cross the gut lining even if you don't have leaky gut.[52] And once it's in your body, WGA binds to sialic acid, part of the sugary coating on many of our cellular surfaces, like blood vessels, synovial surfaces in joints, the BBB, the myelin sheath that encases and insulates your nerves, and even the surface of your eyeballs, leading to inflammation and disease.[53]

Another type of lectins to look out for are called aquaporins, which control the process by which plants "breathe" through the pores in their leaves. Humans also have aquaporins along the gut wall, the BBB, and the lining around our nerves called the myelin sheath that are nearly identical to plant aquaporins. If you develop antibodies to certain plant aquaporins, your immune system might erroneously attack your own aquaporins. This can cause leaky gut, neuroinflammation, and even multiple sclerosis (MS).

In addition to whole grains, aquaporins and other lectins are present in most nightshade vegetables, including tobacco, bell peppers, tomatoes, potatoes, soybeans, plus corn and spinach. Lectins are also found in beans, lentils, pseudo-grains (like amaranth, quinoa, and buckwheat), peanuts, cashews, and chia seeds.

The good news is that most lectin-containing foods can be consumed after being pressure-cooked—or fermented! During the fermentation process, bacteria eat the vast majority of those nasty lectins, making them safer for you and the gut buddies inside of you.

Don't: Eat Too Much Protein

One of the main problems with diets that rely too heavily on protein is that they lead to an overproduction of insulin-like growth factor 1 (IGF-1), a growth hormone that drives the aging process. IGF-1 levels are controlled by a pathway called the mammalian (or mechanistic) target of rapamycin (mTOR) along with your cells and communicate with each other about how much energy is currently available. When mTOR gets the message that there is plenty of energy, it activates IGF-1, and this causes your cells to grow. When mTOR gets the message that energy is limited, it pulls back on IGF-1, preventing growth.

But, you may be wondering, *What is the problem with cell growth?* Basically, the problem is that we are overfed, and our IGF-1 is pretty much always way too high. This leads everything to grow, including cancerous cells, old cells, faulty cells, and so on. This can speed up brain aging and neurodegeneration, not to mention contribute to many other diseases.

It's therefore a good idea to suppress mTOR at least some of the time. When this happens and IGF-1 goes down, faulty and cancerous cells are repaired or discarded via autophagy so they can't harm you. Then, when mTOR is stimulated again, only healthy cells have a chance to grow.[54]

It turns out that reducing protein is one of the best ways to limit IGF-1. In fact, a human study showed that long-term calorie restriction does not reduce IGF-1 levels if protein intake remains high. Reducing protein intake is much more effective at lowering IGF-1, even when consuming the same number of calories![55]

Further, eating an excess of animal protein is linked to an increased risk of depression.[56] Specifically, the amino acid proline can contribute to or cause depression. Gelatin and bone broth are the top sources of proline, so be careful when using these. However, even this link comes down to the microbiome—a balanced inner terrain with adequate amounts of the gut buddy *L. plantarum* protects against the depression caused by consuming too much proline![57]

Mary Anne is a self-proclaimed biohacker in her late twenties. It's never too early to start making changes to "stay young," right? Well, not so fast. Mary Anne came to me with irritable bowel syndrome (IBS) and had been told by a gastroenterologist that she also had small intestinal bacterial overgrowth (SIBO) and candida (which is a yeast overgrowth). Maybe it was the IBS or the SIBO or the candida, but Mary Anne had also noticed that her normal flair and joy for life was gone. She had felt depressed recently but couldn't put her finger on why. Her doctor wanted to start her on multiple antibiotics to take care of her problems. As a "biohacker," Mary Anne knew the downsides of the overuse of antibiotics, so she came to me for alternatives.

I did my usual blood tests and a gut microbiome test. Mary Anne had leaky gut and her vitamin D was low, but the good news was that there was no evidence to support the fact that she had SIBO or candida. Then why was she so depressed? Was it just her leaky gut? Maybe, but . . .

As part of my program, I ask many of my patients to give me a food diary for a week or two, and in Mary Anne's case, the results leapt off the pages! Within the last few months, she had jumped on the bone broth and collagen and organ meats craze praised by many biohackers as the fountain of youth. She was consuming several cups of beef bone broth a day, scoops of collagen several times a day, liver on a regular basis, and lots of grass-fed, grass-finished cuts of beef. When I pointed this out to her, she noticed (now that she thought about it) the more of this hacking she did, the worse her mood had become.

I showed Mary Anne the study mentioned above that describes how the amino acid proline causes alterations to the microbiome, bacterial translocation (due to leaky gut), and GABA suppression. This often leads to depression (particularly in women).

Out the bone broth and other high-proline foods went! In went *L. plantarum*, a GABA-producing probiotic, and the rest of the Gut-Brain Paradox Program, and within weeks, the depression lifted, along with her symptoms of IBS, SIBO, and candida—all without antibiotics. She just needed to hack her biohacking.

Don't: Avoid Nicotine at All Costs

I'm ending this section with a fun, albeit controversial, rule. While smoking is terrible for your health, nicotine is actually very good for the brain. It stimulates neurogenesis, inhibits neuroinflammation, and protects the brain from oxidative stress and aging.[58]

In those with the APOE 4 gene that predisposes them to Alzheimer's, nicotine enhances memory.[59] Nicotine can even prevent the survival and proliferation of glioma cancer cells in brain tumors[60] and appears to protect against schizophrenia. When pregnant rats were injected with LPSs during pregnancy, it induced schizophrenia-like behavior in their offspring, and nicotine reversed the behavior.[61] Further, in children with autism, aggressive behaviors were ameliorated by transdermal nicotine administration.[62]

In addition, nicotine is a powerful mitochondrial uncoupler![63] With its neuroprotective effects, it's no surprise that a study showed smoking can reduce the incidence of Parkinson's disease by 30 percent[64] or that other studies show a correlation between smoking and a reduced risk of dementia.[65]

However, I am not suggesting that you start a cigarette—or any other smoking or vaping—habit. Smoking causes oxidative stress that damages cells and can cause brain aging, basically undoing any of the benefits of nicotine and potentially worse. Of course, the market is now flooded with nicotine lozenges, mints, patches, and other products that can help you get the brain benefits of nicotine without the downsides of smoking. But don't forget that nicotine itself may be addictive. As I write this, the jury is out. And your gut bugs love it![66] So, for now, proceed here with caution.

And with that said, let's move on to the finer details of the Gut-Brain Paradox Program.

CHAPTER 10

WHEN AND WHAT TO EAT ON THE GUT-BRAIN PARADOX PROGRAM(S)

Now that you have a good sense of how to approach this eating plan in general, let's get more specific about what you're going to eat—and, just as important, when.

TIME-RESTRICTED EATING

As with all my food protocols, one of the cornerstones of the Gut-Brain Paradox Program is adjusting your eating schedule to reduce neuroinflammation and restore and nourish a healthy inner terrain. You are probably familiar with the concept of intermittent fasting, which is essentially shortening the period of the day during which you eat. This is also called time-restricted or time-controlled eating. Whatever you want to call it, this way of eating is best for your gut and your brain, and so it is the way of eating that we are going to adopt!

Time-restricted eating benefits the inner terrain in many ways, and the result is that it prevents brain aging and neurodegeneration.[1] Temporary food restriction reduces inflammation stemming from endotoxemia.[2] In mice, it even counters the harmful impact of a Western "cafeteria diet," which includes increased LPSs and neurological changes.[3]

Also in mice, periods of fasting reduce biomarkers of biological age![4] When paired with probiotics, time-restricted eating can drastically improve endotoxemia.[5] Time-restricted eating also assists SCFA production and causes your body to produce ketones, which uncouple mitochondria while strengthening the gut wall.

In addition, this method of eating gives your gut buddies and the wall of your gut a much-needed break. Digestion is a lot of work, and when your gut buddies are digesting food all day long, it creates wear and tear along the gut wall. By restricting the time during which your gut buddies are busy digesting your food for you, you are strengthening (and, more important, repairing) your gut wall, thus reducing neuroinflammation.

Rest assured that this eating method is not as difficult or painful as it may initially seem, and my program will help you mentally and physically gradually adjust to a shortened eating window, the hours during which you will eat in a day. We'll start off with an eating window of twelve hours out of a twenty-four-hour day and shorten it little by little week by week until you get to an eating window of six to eight hours each day, while refraining from eating at least three hours before bedtime. Plus, you only need to maintain this schedule Monday through Friday. Over the weekend, you can eat whenever works best for you.

Weekly Schedule

It will take you five weeks to work your way to the ideal eating window, with each week breaking down as follows:

- **Week 1:** This week is simple. Start breakfast at 8 a.m. and finish your last meal of the day by 7 p.m., Monday through Friday.

Once the weekend comes around, you can be more flexible, but please try to avoid eating too close to bedtime, and make sure you're following the rest of the protocol's dos and don'ts and food lists.

Keep in mind, I generally use 7 p.m. as the end of the day's eating window because this gives most people time for that three-hour eating break before bedtime. But you can adjust this schedule to suit your lifestyle as long as you are working toward a shortened (six- to eight-hour) eating window. For instance, to start, you can eat breakfast at 6 a.m. and stop eating at 5 p.m. if that works better for you.

- **Week 2:** This week is quite similar to week 1, except that this week you will delay your first meal of the day by one hour and eat breakfast at 9 a.m. Once again, stop eating at 7 p.m., feel free to be more flexible on the weekend, and follow the rest of the protocol.
- **Weeks 3 through 5:** Each week, push breakfast back by another hour. This means that during week 3, breakfast will be at 10 a.m. During week 4, breakfast will be at 11 a.m., and during week 5, breakfast will be at noon. That's it! You've successfully achieved a seven-hour eating window and are on your way to a healed gut and a sharper, healthier brain.

Of course, I do acknowledge that the first two weeks of this schedule in particular can be challenging, but your gut buddies and I truly hope you will stick with it. And once they are better able to thrive, your gut buddies will send messages to your brain telling you to keep it up! This will cut down on your hunger and make it feel much easier to keep going.

In the meantime, there are a few things you can do to make it easier. First, make sure that you're drinking plenty of water. Staying hydrated will help keep hunger at bay. Try starting each day with my favorite "healthy soda," San Pellegrino sparkling water with a splash of balsamic vinegar.

Why vinegar? It's an important SCFA postbiotic that serves as a scaffold to build more important SCFAs like butyrate. And feel free to indulge in black coffee or green or black tea. These drinks have plenty of polyphenols, plus caffeine also uncouples mitochondria. Again, just make sure your coffee is black, or choose one of the many keto MCT creamers on the market, which don't count as food during the fasting window.

Speaking of MCT, you can try taking a spoonful of MCT oil (preferably the C8 or C10 varieties, which are more ketogenic) three times a day to keep hunger away. Or take a few capsules or a scoop of ketone salts or esters. More on ketones in the next chapter, but for now, know that these supplements tell your mitochondria to uncouple.

Another option is to mix a scoop of prebiotic fiber powder like psyllium or ground flaxseed into some water. Because you can't digest prebiotic fiber, this won't break your fast, but it will feed your gut buddies. They will then send a signal to your brain saying that you are full!

If all else fails and you are feeling too hungry, weak, cranky, or tired to keep going, simply slow down and go back to the previous week's eating schedule. It's better to take more than five weeks to get to a six- to eight-hour window than it is to give up entirely! Once you adjust and begin feeling better, start slowly pushing breakfast back by another hour.

The "Ramadan" Schedule

During Ramadan, Muslims fast from dawn to sunset. Most families get up early and eat a small breakfast before the sun rises and then abstain from food or drink until having their main meal of the day after sunset. For reasons separate from religion, this type of schedule works better for some of my patients than the one outlined above and has many of the same benefits.

By fasting for twelve hours during the day and then again for eight hours overnight, those who follow Ramadan traditions are essentially fasting for twenty hours across a twenty-four-hour period. If you like, try eating an early breakfast, skipping lunch, and then waiting for dinner as your last meal of the day. As always, just make sure you are following the other aspects of the protocol.

One Meal a Day

Another option that works well for some people is the one-meal-a-day (OMAD) plan. This will likely be a good choice for you only if you already practice intermittent fasting. If this is you or if you find week 1 of the program to be a breeze, you can speed up your progress by eating a single meal each day. I personally follow this protocol from January to June each year. During this time of year, when food resources would naturally be scarcer in a more primitive society, I eat just one meal between 5 p.m. and 7 p.m. and fast for the rest of the day.

If you're not already practicing intermittent fasting and want to give OMAD a try, I recommend following the five-week protocol first. Then during week 6, you can push your first meal back to 1 p.m. Week 7, make it 2 p.m. And so on and so forth until you reach the two-hour OMAD window by week 11. You can eat your one meal at whatever time you like, as long as it's not too close to bedtime.

While this schedule is certainly beneficial, I don't recommend that you follow it consistently for the long term. Make sure to alternate between OMAD and other time-restricted eating schedules. Perhaps try OMAD Monday through Friday and then have two or three meals a day over the weekend, or do what I do and only follow this plan for part of the year.

Okay, now that we've covered the "when," let's move on to the "what"!

THE GUT-BRAIN PARADOX FOOD LISTS

To take all the guesswork out of the program, here are complete lists of foods, broken down into just two categories. "Yes, Please" foods are the ones you should feel free to indulge in during your eating window. The "No, Thank You" foods are the ones that you should make sure to avoid.

Yes, Please: Postbiotic-Boosting Foods

Cruciferous Vegetables

Arugula

Bok choy

Broccoli

Brussels sprouts

Cabbage, green and red

Cauliflower

Collards

Kale

Kimchi

Kohlrabi

Napa cabbage

Sauerkraut (raw or canned)

Swiss chard

Watercress

Other Postbiotic-Boosting Vegetables

Artichokes

Asparagus

Bamboo shoots

Basil

Beets (raw)

Carrot greens

Carrots (raw)

Celery

Chicory

Chives

Cilantro

Daikon radishes

Endive

Escarole

Fiddlehead ferns

Frisée

Garlic

Garlic scapes

Ginger

Hearts of palm

Horseradish

Jerusalem artichokes (sunchokes)

Leeks

Lemongrass

Mesclun

Mint

Mizuna

Mushrooms (always cook white button mushrooms)

Mustard greens

Nopales (cactus paddles; if you can't find them locally, buy online)

Okra

Onions

Parsley

Parsnips

Perilla

Puntarelle (an Italian chicory)

Purslane

Radicchio

Radishes

Red- and green-leaf lettuces

Romaine lettuce

Rutabaga

Scallions

Sea vegetables

Seaweed and algae

Shallots

Spinach (warning—contains an aquaporin lectin)

Treviso (a cousin of radicchio)

Water chestnuts

Fruits That Act Like Fats

Avocado (up to a whole one per day)

Olives, all types

Oils

Avocado oil

Black seed oil

Canola oil (non-GMO, organic only!)

Coconut oil

Cod liver oil (the lemon and orange flavors have no fish taste)

Flaxseed oil (high-lignan)

Macadamia oil (omega-7)

MCT oil

Olive oil (organic, extra-virgin, first cold-pressed)

Perilla oil (lots of ALA and rosmarinic acid, both uncouplers)

Red palm oil

Rice bran oil

Sesame oil, plain and toasted

Walnut oil

Nuts and Seeds (Up to ½ cup per day)

Almonds (only blanched or Marcona)

Barùkas (or baru) nuts

Basil seeds

Brazil nuts (in limited quantities)

Chestnuts

Coconut meat (but not coconut water)

Coconut milk/cream (unsweetened full-fat canned)

Coconut milk (unsweetened dairy substitute)

Duckweed powder

Flaxseeds (ground fresh)

Hazelnuts

Hemp protein powder

Hemp seeds

Macadamia nuts

Milkadamia creamer (unsweetened and not the milk, which
usually contains pea protein)

Nut butters (if almond butter, preferably made with blanched
almonds, as almond skins contain lectins)

Pecans

Pili nuts

Pine nuts

Pistachios

Psyllium seeds/psyllium husk powder

Sacha inchi seeds

Sesame seeds

Tahini

Walnuts

Processed Resistant Starches

Can be eaten every day in limited quantities; those with prediabetes or diabetes should consume only once a week on average.

Barely Bread breads and bagels (only those without raisins)

Bread SRSLY sourdough non-lectin bread and rice-free sourdough rolls

Cappello's fettuccine and other pasta

Crepini egg wraps

Fullove Foods keto hemp and linseed bread

Julian Bakery Paleo wraps (made with coconut flour), Paleo thin bread, almond bread, sandwich bread, coconut bread

Lovebird Cereals (unsweetened only)

ONANA Tortillas

Positively Plantain tortillas

Real Coconut coconut and cassava flour tortillas and chips

Siete chips (be careful here; a couple of my "canaries" have reacted to the small amount of chia seeds in the chips) and tortillas (only those made with cassava and coconut flour or almond flour)

Terra cassava, taro, and plantain chips (be careful; many of my patients eat too many!)

Thrive Market organic coconut flakes

Tia Lupita grain-free cactus tortillas

Trader Joe's jicama wraps, plantain chips

Resistant Starches

> *Eat in moderation. People with diabetes and prediabetes should initially limit these foods.*

Baobab fruit

Cassava (tapioca)

Celery root (celeriac)

Glucomannan (konjac root)

Green bananas

Green mangoes

Green papayas

Green plantains

Gundry MD Popped Superfood Crisps

Jicama

Millet

Parsnips

Persimmon

Rutabaga

Sorghum

Sweet potatoes or yams

Taro root

Tiger nuts

Turnips

Yucca

"Foodles" (Acceptable "Noodles")
   *Note—diabetic, prediabetic, and insulin-resistant people should use
   these with extreme moderation except for konjac-based noodles and
   rice or hearts of palm noodles or rice.*

Big Green millet and sorghum pastas

Edison Grainery sorghum pasta

Gundry MD konjac shirataki noodles

Gundry MD sorghum spaghetti

Jovial cassava pastas

Kelp noodles

Konjac noodles

Miracle Noodle kanten pasta

Miracle Rice

Natural Heaven hearts of palm spaghetti and lasagna
   noodles

Palmini hearts of palm noodles

Shirataki noodles

Slimdown360 sweet potato pasta elbow macaroni

Trader Joe's cauliflower gnocchi

Wild-Caught Seafood (4 ounces per day)
   Note: Use with caution due to microplastic content.

Alaskan salmon (very few microplastics)

Anchovies

Calamari/squid

Canned tuna

Clams

Cod

Crab

Freshwater bass

Halibut

Hawaiian fish, including mahi-mahi, ono, and opah

Lake Superior whitefish

Lobster

Mussels

Oysters

Sardines

Scallops

Shrimp (wild only)

Steelhead

Trout

Pastured Poultry (4 ounces per day)

Chicken

Duck

Game birds (pheasant, grouse, dove, quail)

Goose

Heritage or pastured turkey

Ostrich

Pastured chicken or turkey jerky (low-sugar versions)

Pastured or omega-3 eggs (up to 4 daily)

Meat (4 ounces per week)
  100 percent grass-fed and grass-finished; see the previous chapter.

  Beef

  Bison

  Boar

  Elk

  Pork (humanely raised, including prosciutto, Ibérico ham, and
    Cinco Jotas ham)

  Traditionally fermented sausages (Hint: Look for the words
    "lactic acid cultures." Good news: They have Neu5Gc—
    see page 169.)

Plant-Based Proteins and "Meats"

  Duckweed powder

  Flaxseed protein powder

  Gundry MD ProPlant protein shakes

  Hemp protein powder

  Hemp tofu

  Hilary's root veggie burger (PlantPlus Foods)

  Just Eggs (ju.st)

  Perfect Day vegan whey and casein

  Pressure-cooked lentils and other legumes (canned, such as Eden
    or Jovial brand) or dried, soaked, then pressure-cooked (use an
    Instant Pot)

  Protein isolates of and/or hydrolyzed pea, soy, or other similar bean powders (not the same as regular pea protein, soy protein, lentil protein, chickpea protein—buyer beware!)

Quorn products: only meatless pieces, meatless grounds, meatless steak-style strips, meatless fillets, meatless roast (avoid all others, as they contain lectins/gluten)

Textured vegetable protein (TVP)

Polyphenol-Rich Fruits

Limit to one small serving on weekends and only when that fruit is in season, or unlimited with "reverse juicing." This entails juicing your fruit and then instead of throwing out the pulp and drinking the juice, throw out the juice, which contains most of the fructose, and eat the pulp, instead!

Best options are pomegranate and passion fruit seeds, followed by raspberries, blackberries, strawberries, then blueberries, grapefruit, pixie tangerines, and kiwis. (Eat the skin for more polyphenols.)

Apples

Apricots

Blackberries

Blueberries

Cherries

Citrus, all types (no juices)

Cranberries (fresh)

Guava

Kiwis

Nectarines

Papayas

Passion fruit

Peaches

Pears, crispy (Anjou, Bosc, Comice)

Persimmons

Plums

Pomegranates

Raspberries

Starfruit

Strawberries

Dairy Products and Replacements

Aged cheeses from Switzerland

Aged "raw" French/Italian cheeses

Buffalo butter (available at Trader Joe's)

Buffalo mozzarella: mozzarella di bufala (Italy), Buf Creamery (Uruguay)

Coconut yogurt (plain)

French/Italian butter (limit)

Ghee (grass-fed; limit)

Goat's and sheep's milk kefir (plain)

Goat ghee (limit)

Goat milk cream flakes (Mt. Capra)

Goat's milk cheeses: feta, Brie, mozzarella, cheddar

Goat's milk yogurt (plain)

Kite Hill ricotta

Lavva plant-based yogurt

Organic heavy cream

Organic sour cream

Parmigiano-Reggiano cheese

Sheep's milk cheeses: Pecorino Romano, Pecorino Sardo, feta, Manchego

Sheep's milk yogurt (plain)

So Delicious vegan mozzarella, cream cheese

Herbs, Seasonings, and Condiments

Avocado mayonnaise

Coconut aminos

Fish sauce

Herbs and spices (all except red pepper flakes)

Lea & Perrins Worcestershire sauce (no other brands)

MCT oil mayonnaise

Miso paste

Mustard

Nutritional yeast

Pure vanilla extract

R's KOSO, other KOSOs

Sea salt (iodized)

Tahini

Vinegars (apple cider vinegar, BLiS vinegars, Sideyard Shrubs vinegars, and others)

Wasabi

Flours

Almond (blanched, not almond meal)

Arrowroot

Cassava

Chestnut

Coconut

Coffee fruit

Grape seed

Green banana

Hazelnut

Millet

Sesame (and seeds)

Sorghum

Sweet potato

Tiger nut

Sweeteners

Allulose (By far the best option! Look for non-GMO)

Erythritol (Not evil, as some think; Swerve is my favorite, as it also contains oligosaccharides)

Inulin (Just Like Sugar is a great brand)

Local honey and/or Manuka honey (very limited amounts!)

Monk fruit (luo han guo; the Nutresse brand is good)

Stevia (SweetLeaf is my favorite; also contains inulin)

Xylitol

Yacon syrup (Sunfood Sweet Yacon Syrup is available on Amazon)

Chocolate and Frozen Desserts

Coconut milk dairy-free frozen desserts (the So Delicious blue label, which contains only 1 gram of sugar; but be careful: It may contain pea protein)

Dark chocolate, unsweetened, 72% cacao or greater (1 ounce per day)

Enlightened ice cream

Keto ice cream: chocolate, mint chip, sea salt caramel

Killer Creamery ice cream: Chilla in Vanilla, Caramels Back, and No Judge Mint

Mammoth Creameries: vanilla bean

Natural (non-Dutched) cocoa powder, unsweetened

Nick's vegan ice cream

Rebel ice cream: butter pecan, raspberry, salted caramel, strawberry, vanilla

Simple Truth ice cream: butter pecan and chocolate chip

Beverages

Please limit alcohol consumption to no more than 6 ounces of wine or 1 ounce of dark spirits per day.

Champagne (up to 6 ounces per day)

Coffee

Dark spirits (up to 1 ounce per day)

Hydrogen water

KeVita brand low-sugar kombucha (coconut, coconut mojito, for example), other low-sugar kombuchas

Red wine (6 ounces per day)

Reverse osmosis filtered water (AquatruPro.com)

San Pellegrino or Acqua Panna water, or other Italian sparkling waters

Tea (all types)

No, Thank You

Refined, Starchy Foods

Bread

Cereal

Cookies

Crackers

Pasta

Pastries

Potato chips

Potatoes

Rice

Tortillas

Wheat flour

Whole wheat flour

Grains, Sprouted Grains, Pseudograins, and Grasses

Barley (cannot pressure-cook)

Barley grass

Brown rice

Buckwheat

Bulgur

Corn

Corn products

Corn syrup

Einkorn

Kamut

Kasha

Oats (cannot pressure-cook)

Popcorn

Quinoa

Rye (cannot pressure-cook)

Spelt

Wheat (pressure cooking does not remove lectins from any form
of wheat)

Wheatgrass

White rice (except pressure-cooked white basmati rice from
India, which is high in resistant starch; American white
basmati is not)

Wild rice

Sugar and Sweeteners

Agave nectar

Coconut sugar

Diet drinks

Granulated sugar (even organic cane sugar)

Maltodextrin

NutraSweet (aspartame)

Splenda

Sweet'N Low (saccharin)

Sweet One and Sunett (acesulfame-K)

Vegetables

Most of these can be made safe foods with pressure cooking; marked with an (*).

Beans* (all, including sprouts)

Chickpeas* (including as hummus)

Edamame*

Green/string beans*

Legumes*

Lentils (all)*

Pea protein (unless pea protein isolate or hydrolysate)

Peas*

Soy*

Soy protein (unless soy protein isolate or hydrolysate)

Sugar snap peas

Tofu*

Nuts and Seeds

Almonds, unblanched

Cashews

Chia seeds

Peanuts

Pumpkin seeds

Sunflower seeds

Fruits

Some we call vegetables.

Bell peppers

Chili peppers

Cucumbers

Eggplant

Goji berries

Melons (any kind)

Pumpkins

Squash (any kind)

Tomatillos

Tomatoes

Zucchini

Milk Products that Contain A1 Beta-Casein

Butter (even grass-fed), unless from A2 cows, sheep, goats, or buffalo

Cottage cheese

Cow's milk

Cow's milk cheese from American cows

Frozen yogurt

Ice cream (most)

Kefir from American cows

Ricotta (unless imported from Italy)

Yogurt (cow's milk; including Greek yogurt)

Oils

Corn

Cottonseed

Grape seed

Partially hydrogenated oils (all)

Peanut

Safflower

Soy

Sunflower

"Vegetable"

Herbs and Seasonings

Ketchup

Mayonnaise (unless MCT or avocado)

Red pepper flakes

Soy sauce

Steak sauce

Worcestershire sauce (unless Lea & Perrins)

* * *

For most people, this plan will work well to restore a flourishing inner terrain and a robust gut wall, thereby reducing or downright eliminating neuroinflammation and all the problems it can cause. However, there are those who will need a bit more help to heal their brains from a severe mental health issue or addiction. The following chapter will provide a protocol that is especially designed to help with these treatment-resistant conditions. There is so much hope! So, let's go get some.

CHAPTER 11

THE CHICKEN AND THE SEA (THE GUT-BRAIN PARADOX MODIFIED CARNIVORE DIET)

When I published *The Plant Paradox* in 2017, warning readers about and documenting the harmful effects of plant defense compounds like lectins, I started an uproar about the relationship of plants to animals that continues to this day. When carried to the extreme, the science suggests that any plant defense compound, like lectins, oxalates, phytates, indoles, glucosinolates, tannins, and even polyphenols could be potentially harmful.[1]

It turns out these plant compounds are most harmful in people who do not have a strong inner terrain and/or have intestinal permeability. As just one example, it has recently been discovered that individuals who are sensitive to oxalates and experience pain and/or oxalate kidney stone formation lack the gut buddy *Oxalobacter formigenes* and other lactic acid bacteria. These gut buddies would normally degrade (eat) oxalic acid, rendering them harmless.[2]

To put it another way, animals and their microbiomes evolved to detoxify plant defense compounds in the gut before they could impact the animal. Therefore, an animal or individual with a beautifully intact and diverse terrain would be immune (or protected) from these plant defense compounds and would never "notice" them as a problem. On the other hand, if the inner terrain was decimated or not as robust as normal and/or the gut wall was porous, this animal or individual might be set up to be negatively affected by those plant defense compounds. In other words, they are a sitting duck.

If you read *The Plant Paradox*, you may recall that I referred to these extremely sensitive individuals as my "canaries," as in, "canaries in a coal mine." They could spot troublemaking foods right away because they experienced symptoms immediately when consuming them. In these same folks, subsequent food sensitivity tests were remarkably spot-on.

So, where is this all going? In some individuals with a desert wasteland for an inner terrain and an extremely porous gut wall, any plant material may be intolerable. Indeed, many proponents of the so-called carnivore diet (which is a misnomer, in my opinion) have called me the true father of the carnivore diet.

Sadly, as you now know, most carnivore diets are based primarily on beef, with the addition of lamb, pork, bison, organ meats, and maybe poultry and fish. But it's mainly beef, maybe with the proviso of grass-fed and grass-finished. Yet, with the exception of poultry and fish, all of these "carnivore" meats cause and/or perpetuate leaky gut, leaky brain, and neuroinflammation, for reasons I will discuss in a moment.[3,4]

Finally, to put the nail in the carnivore-diet coffin, in a recent human trial looking at this type of diet, both insulin resistance and inflammation increased, despite (and this is the important part) the fact that the participants felt better.[5] I have seen the exact same thing in my own patients. They felt better, but their inflammation numbers looked worse. Can you guess why? Go to the head of your class if you said their blood levels of LPSs went up! And what do LPSs ride on to get across the wall of the gut? Fat. The saturated fat in all that fatty beef.

But they felt better! So, what to do? With these facts in mind, I designed this program to ameliorate the downsides of the carnivore diet while providing the mood and pain-relieving benefits that my patients reported. And the good news is, it worked!

This program is specifically designed for those with challenging mental health, addiction, abdominal, and/or other pain issues. However, it can also jump-start your progress no matter where you're starting from. Either way, this is meant to be a relatively short-term reset, not a long-term dietary plan. I recommend sticking with it for at least six weeks and up to three months. This is the length of time it generally takes for my patients to start seeing big positive changes. Then, you can switch over to the regular Gut-Brain Paradox Program.

Even if this program is not for you, the information about why it is so beneficial is certainly useful. So, I encourage you to read on even if you think it may not apply to you. This program draws on the brain benefits of a traditional ketogenic (keto) diet while helping you avoid the negative brain (and gut) consequences that sadly accompany many ketogenic or carnivore diet plans. I spent an entire previous book, *Unlocking the Keto Code*, profiling why many "keto diets" are detrimental to building a balanced inner terrain. If you want more details, you will find them there.

KETO AND THE BRAIN

So, what exactly are the brain benefits of the keto diet? Good question. We've actually known that ketones are good for the brain for well over a century, long before we understood how they worked. This was discovered when doctors found that when children with epilepsy were given a diet consisting of 80 percent fat, 10 percent protein, and 10 percent carbohydrates, the frequency and severity of their seizures was significantly reduced. For many children, this protocol stopped the seizures altogether.

The only other dietary intervention that worked almost as well as this diet was full fasting. Children who consumed only water for eighteen to twenty-four hours at a time often saw marked neurologic improvements. Of course, the keto diet (which was not called that at the time) was a far more realistic option for children than a full fast.

Then, in 1921, an endocrinologist at Northwestern University named Rollin Turner Woodyatt first learned that the liver produces compounds called ketones when we are either starved or when consuming a diet rich in fats but restrictive in protein and carbohydrates. The liver does this by picking up free fatty acids—lipids that come directly from the fats we produce and store in our fat cells—and converting them into ketones. Ketones are small enough molecules (unlike free fatty acids) that they can pass through the BBB and be used by the brain as an "emergency" fuel source when glucose levels are low.

Within a year, a researcher at the Mayo Clinic named Dr. Russell Wilder developed a high-fat, low-carb diet that he called a "ketogenic diet." He successfully used this diet to treat children with epilepsy. In fact, until antiseizure medications came along, the ketogenic diet was the standard of treatment for childhood epilepsy.

So, why is a ketogenic diet such a good treatment for epilepsy, and, assuming you are not suffering from seizures, what does this mean for you? First off, ketones work by signaling your mitochondria to uncouple and produce more of themselves through mitogenesis. This strengthens the gut wall and the BBB, reducing neuroinflammation.

Ketones do the same thing for the mitochondria in your brain. And, just to be clear, your brain neurons do not like to use ketones as their preferred fuel! Instead, when ketones activate uncoupling proteins in neurons, it increases the heat in those brain cells. It's this effect that actually increases the neurons' individual function![6]

Okay, so far so good. But again, you are probably not suffering from seizures, so what benefits for brain health and mood and addiction could accrue by generating ketones? If you guessed diminishing neuroinflammation, you are spot-on! Indeed, both human and animal studies show that a supplemented ketogenic diet (or one that adds additional ketone-making MCT oils to a low-carb diet) dramatically protects microglial activation and improves depressive and additive behaviors.[7,8]

You may be wondering about the butyrate production that happens when your gut buddies ferment soluble fiber from plants. Isn't that critical, as well? Don't be shocked, but amino acids from proteins can also be fermented into SCFAs, including butyrate.[9] This means that a high-protein ketogenic diet strengthens the gut wall and the BBB through both uncoupling and increased butyrate production! A stronger gut wall and a reinforced BBB plus less neuroinflammation plus better functioning neurons is a powerful formula.

A brand-new study released in the spring of 2024 showed that patients with schizophrenia had an average improvement in symptoms of 32 percent after four months on a ketogenic diet, and 69 percent of patients with bipolar disorder also saw a significant improvement in symptoms.[10] In another human study of a keto diet in adults with severe, persistent mental illnesses such as major depressive disorder, bipolar disorder, and schizoaffective disorder, patients saw substantial improvements in depression and psychosis symptoms.[11]

Further, when older adults adopted a modified keto diet featuring more fish and poultry than red meat (spoiler alert: That's what this particular program is all about), it altered the makeup of their microbiomes in ways that increased butyrate production and impacted cognitive function.[12] And even in the general population, a keto diet is associated with improved mental and emotional well-being and a reduction in depression and anxiety.[13]

What about addiction? Studies show that a keto diet can even help reduce withdrawal symptoms to a variety of addictive substances.[14] But of course! Because this diet changes the messages being sent from the gut to the brain, these messages no longer tell the person suffering from addiction that they need to continue ingesting that addictive substance.

THE DOWNSIDES OF KETO FOR THE BRAIN

This all sounds pretty good, doesn't it? So, what are the negative consequences of a traditional keto or carnivore diet? Another excellent question.

The two main downsides of a traditional keto diet both stem from its overreliance on animal protein—in particular beef. For one thing, beef is full of saturated fat. As you read earlier, saturated fat leads to metabolic disturbances that can cause cognitive decline. In addition, there is a direct association between saturated fat intake and LPS concentrations in the blood.[15]

Equally if not more problematic is a specific sugar molecule along the gut wall and BBB called Neu5Gc in beef, pork, lamb, and bison. These, of course, are all common foods on a typical ketogenic diet. Meanwhile, humans, fish, and chicken have an incredibly similar but not quite identical sugar molecule along our gut walls and BBB called Neu5Ac.

The difference between Neu5Gc and Neu5Ac is literally only one oxygen molecule. Yet, this still means that Neu5Gc is foreign to our immune system. When we eat foods containing Neu5Gc, we make antibodies to Neu5Gc. The more Neu5Gc-containing foods we eat, the more antibodies to it we make.[16] This is the same way that food sensitivities develop.

The brain actively keeps Neu5Gc out to protect itself.[17] But when you eat foods containing Neu5Gc, it is rapidly absorbed in your small intestine and can be incorporated into your gut wall and even the BBB, replacing the Neu5Ac that is supposed to be there. Again, the more foods with Neu5Gc you eat, the more this happens, all while your body produces more and more antibodies to Neu5Gc. The immune system them starts to attack that Neu5Gc along the gut wall and BBB, leading to an inflamed gut wall and brain.[18,19]

Normally, I recommend that my patients eat a lot less red meat but still enjoy grass-fed, grass-finished beef and pastured lamb or pork on rare occasions while upping their intake of Neu5Ac foods. Why? Because there is evidence that Neu5Ac can protect you from the impact of Neu5Gc.[20] And in the regular Gut-Brain Paradox Program I am also including cured or fermented meats. When meats are fermented, bacteria eat much of the Neu5Gc, so it cannot hurt you![21] Those good old gut buddies really do have your best interest at heart.

This is another reason why fermented dairy is also allowed on both programs but not unfermented dairy. Milk from Neu5Gc animals also contains this sugar molecule. But fermentation gets rid of the vast majority of Neu5Gc in dairy products.

However, for those suffering from significant brain ailments, I am recommending that you avoid red meat entirely on this program, except occasional properly fermented (dry-aged or wet-aged for at least four weeks) grass-fed, grass-finished beef, prosciutto, or 5J hams, and sausages only made by traditional fermentation. Even better, give yourself a chance to see how it feels to eliminate red meat entirely. I am confident that the results will speak for themselves.

Want more motivation? A study published in May 2024 that followed four thousand American women listed the foods most associated with accelerated aging. Here's the rundown: eggs, organ meats (no surprise—beef liver has the highest Neu5Gc content of any food), sausages, cheese, legumes, starchy vegetables, added sugar, and lunch meats. And the foods most associated with decelerated aging were poultry, nuts, peaches, nectarines, plums (all high in polyphenol content), and solid fats (like butter and fermented cheeses).[22] Sounds like this program, doesn't it?

George was a confirmed gluten-free vegetarian (almost vegan) for forty years. But when he started working with me, his gut was a mess, and he was not the happiest person I had ever met. Let's just say that he had to plan any activity away from his house carefully for pit stops along the way.

George was convinced that gluten was his problem, which he couldn't avoid thanks to hidden gluten lurking everywhere. His blood work did show anti-gluten antibodies, as well as bad leaky gut and LPSs. We went on for years modifying his essentially gluten-free and now lectin-free diet, to really no avail. He religiously took many polyphenol-containing supplements and drank coffee for its polyphenol content. Yet nothing changed.

Finally, George's spouse and I had enough. We pleaded with him to try this program that you are reading about here, which means stopping all plants, including all spices except for salt, and all supplements except vitamin D_3 and fish oil. And this confirmed vegan/vegetarian dove off the deep end! Chicken skin, chicken livers, pastured chicken, fish, oysters, shrimp, fermented sausages, and occasional dry-aged beef suddenly became his only foods. What a shock to the system, right?

Wrong! Within two weeks, George was on the phone telling me that he had his first normal bowel movement in forty years. His workout energy skyrocketed, and his spouse reported that his depression and grouchiness disappeared. After three months, we started slowly titrating back in polyphenols, which he was able to tolerate, but any attempt to bring back plant material/fiber was met with bowel changes. Every single time. But his blood work? No more leaky gut and no more LPSs.

George has now had two years of happiness on this program, and you can, too.

THE CHICKEN AND THE SEA FOODS

So, how can you benefit from the brain-boosting effects of a keto diet without the harmful impact of non-fermented red meat? As the name suggests, this plan is a simple approach to keto that focuses completely on any poultry and seafood that's either baked, sauteed, poached, stir-fried, or grilled with perilla oil or Ahiflower oil, plus fermented dairy products. That's right—it's the Pescatarian-Fowl-Fermented-Dairy-Fermented-Carnivore Diet! (Chicken and the Sea seemed catchier.)

On this program, there's no fruit, no veggies, no salads, no tubers, no "safe" pastas, none of those. And the only oils allowed are perilla oil and Ahiflower oil, owing to their unique ability to reduce LPSs. This may sound extreme, but I'm telling you that my patients love this program once they get used to it—especially after they start to see results.

On this program, it's especially important to buy the highest-quality products you can find and afford. Make sure your poultry is pasture-raised. There are more and more providers on the Internet and appearing at stores like Whole Foods and Trader Joe's, as well. Look for names like "heritage" chicken or duck. And make sure your seafood is wild (not farmed). Factory-farmed chicken and farmed seafood are fed antibiotics, which you consume when you eat these products. Of course, these antibiotics kill off the very gut buddies that you're trying to nurture! And be cautious of even "organic" chicken, as most are fed corn and soybeans.

I do have some exciting news for all you beef and pork lovers. My longtime friend chef Jimmy Schmidt (winner of three James Beard Awards) has developed beef and pork products from grass-fed, grass-finished animals whose meat is fermented to drastically reduce its Neu5Gc content. It is available through his company JRRanchFoods .com. Excuse the expression, but you can now have your cake (or meat) and eat it, too! And, doubters, I have no commercial relationship with Jimmy's company.

You may be wondering, *What about fiber? Don't my gut buddies need fiber in order to thrive and produce butyrate?* Indeed, they do, but don't forget that they can ferment these proteins, as well. Plus, you are going to be consuming animal "fiber" in chicken gristle and skin and the bones of sardines and other small fish, like anchovies and herring.

Finally, this program boosts your consumption of dietary spermidine, a compound shown to improve cognitive performance and mitochondrial function in animal and human trials.[23] Chicken, chicken skin, chicken livers, and aged cheeses are rich sources of spermidine.

You do, however, have to be careful of microplastic content. One study showed that there are now microplastics in nearly all proteins grown or raised in the United States.[24] Horrifying, but the good news is that processed proteins, such as fish sticks, chicken fingers, and even "plant-based" nuggets, had significantly higher levels than unprocessed proteins. Of course, you will not be eating these processed proteins. The proteins tested with the lowest microplastic levels included chicken breast, fresh-caught Alaskan pollack, and fresh-caught white gulf shrimp.

In addition, small fish and shellfish tend to be low in microplastics in general. Wild salmon, oysters, shrimp, sardines, trout, and char are all good choices, as well. I've also recently discovered a fish called mullet, which is high in C15, a fatty acid (pentadecanoic acid) that is highly anti-inflammatory and beneficial for mitochondria.[25] In a human study, the presence of C15 in the blood predicted less heart disease.[26] Besides mullet, you can also get C15 in fermented dairy products.

Remember, this program can be temporary, and it works. I have seen so many patients spend six weeks in a rehabilitation program for addiction only to relapse, but go on to see huge improvements after only six weeks on this version of my Gut-Brain Paradox Program. It accomplishes a total reset of your gut terrain, your gut wall permeability, and your brain's neuroinflammation, all in a remarkably short period of time.

* * *

No matter what condition you are struggling with—or even if you are perfectly healthy and just want to prevent brain degeneration in the future—I have confidence that either of these programs can help you reach your goals. As I said at the beginning of the book, my incredible patients heal themselves every day, simply by following my suggestions. I look forward to adding your success story to this list.

ACKNOWLEDGMENTS

It's always exciting to start writing my next book, and this one was no exception!

With that in mind, I invited my collaborator from my *New York Times* bestseller *The Longevity Paradox*, and our last bestseller together, *Gut Check*, Jodi Lipper, to hop aboard the train again and go for the trifecta! Jodi, we had so much fun, let's do it again soon! And thanks for making my nerdy, professor-speak, thick science accessible to our readers again.

Karen Rinaldi, my longtime overseer at Harper Wave, moved us both over to Harper and took over full-time editing of *The Gut-Brain Paradox*. Thanks for all your valuable insights and contributions.

Naturally, much of what you read on these pages stems from the fact that I continue to see patients six days a week, at the International Heart and Lung Institute in Palm Springs, California, and at the Centers for Restorative Medicine in Palm Springs, Santa Barbara, and Beverly Hills. All of this couldn't happen without the tireless work of Mitsu Killion-Jacobo, my longtime physician assistant and associate director of the International Heart and Lung Institute; Susan Lokken, my right-hand woman of decades as executive assistant and office manager. All this is kept aboveground by my CFO, Joseph Tames, and my friend and attorney, Dave Baron. Thanks to you all!

And thanks to my longtime agent and protector, Shannon Marven, president of Dupree Miller, and her assistant and co-worker Rebecca Silensky.

And a continued shout-out to the hundreds of employees at Gundry MD who have made me, GundryMD.com, *The Dr. Gundry Podcast*, and my YouTube channels an essential source for your health and wellness needs. And a special shout-out (again) to my support team at Gundry MD, headed by Lanee Lee Neil, along with Kate Holzhauer and Jacie Ray, as well as the talented writers and film crew that keep the cutting-edge health information that I discover (thanks to my patients) coming out to you.

Despite the fact that I continue to see patients six days a week, even Saturdays and Sundays, the wait times to see me or Mitsu are very long. Because of that, I've recently launched Gundry Health, its app, and GundryHealth.com, my subscription-based telemedicine service, where you can interact with physician associates, trained by me, to manage and treat all autoimmune conditions, "leaky gut," and IBS, and the neurological disorders you read about here using the same protocol and state-of-the-art blood tests that you read about in this book. Finally, you can now see "me" without the wait! I can't wait to see you there.

Final thanks, of course, goes to my soulmate and wife, Penny, who keeps me grounded and in check while caring for our four dogs, including two rescues. She retired from her twenty-year-old business, Zense, last year, so she has more time to harass me into taking care of myself. Thanks, Penny!

NOTES

INTRODUCTION: LET FOOD BE THY MEDICINE

1. Rees, T., *Plastic Reason* (Berkeley, CA: University of California Press), 2016.
2. Sitar, J. K., "Semilunárnímu zvýšení dopravní nehodovosti" [The effect of the semilunar phase on an increase in traffic accidents], *Casopis Lekaru Ceskych* 1994;133(19): 596–98. Czech. PMID: 7954673.
3. Wehr, T. A., "Bipolar mood cycles and lunar tidal cycles," *Molecular Psychiatry* 2018;23(4): 923–31. https://doi.org/10.1038/mp.2016.263
4. Vyazovskiy, V. V., and Foster, R. G., "Sleep: A biological stimulus from our nearest celestial neighbor?" *Current Biology* 2014;24(12): R557–60. https://doi.org/10.1016/j.cub.2014.05.027
5. Risely, A., Wilhelm, K., Clutton-Brock, T., Manser, M., and Sommer, S., "Diurnal oscillations in gut bacterial load and composition eclipse seasonal and lifetime dynamics in wild meerkats," *Nature Communications* 2021;12: 6017. https://doi.org/10.1038/s41467-021-26298-5

CHAPTER 1: SH*T FOR BRAINS

1. Cavaillon, J. M., and Legout, S., "Louis Pasteur: Between myth and reality," *Biomolecules* 2022;12(4): 596. https://doi.org/10.3390/biom12040596
2. Manchester, K. L., "Louis Pasteur, fermentation, and a rival," *South African Journal of Science* 2007;103(9): 377–80. https://hdl.handle.net/10520/EJC96719
3. Sender, R., Fuchs, S., and Milo, R., "Revised estimates for the number of human and bacteria cells in the body," *PLoS Biology* 2016;14(8): e1002533. https://doi.org/10.1371/journal.pbio.1002533

4. McLaughlin, R. W., Vali, H., Lau, P. C., Palfree, R. G., De Ciccio, A., Sirois, M., Ahmad, D., Villemur, R., Desrosiers, M., and Chan, E. C., "Are there naturally occurring pleomorphic bacteria in the blood of healthy humans?" *Journal of Clinical Microbiology* 2002;40(12): 4771–75. https://doi.org/10.1128/JCM.40.12.4771-4775.2002

5. Arabi, T. Z., Alabdulqader, A. A., Sabbah, B. N., and Ouban, A., "Brain-inhabiting bacteria and neurodegenerative diseases: The 'brain microbiome' theory," *Frontiers in Aging Neuroscience* 2023;15. https://doi.org/10.3389/fnagi.2023.1240945

6. Ironstone, P., "Me, my self, and the multitude: Microbiopolitics of the human microbiome," *European Journal of Social Theory*, 2019;22(3): 325–41. https://doi.org/10.1177/1368431018811330

7. Relman, D. A., "The human microbiome: Ecosystem resilience and health," *Nutrition Reviews* 2012;70(Suppl. 1): S2–9. https://doi.org/10.1111/j.1753-4887.2012.00489.x

8. Lozupone, C. A., Stombaugh, J. I., Gordon, J. I., Jansson, J. K., and Knight, R., "Diversity, stability and resilience of the human gut microbiota," *Nature* 2012;489(7415): 220–30. https://doi.org/10.1038/nature11550

9. Bever, J. D., Westover, K. M., and Antonovics, J., "Incorporating the soil community into plant population dynamics: The utility of the feedback approach," *Journal of Ecology* 1997;85(5): 561–73. https://doi.org/10.2307/2960528

10. Tony-Odigie, A., Dalpke, A. H., Boutin, S., and Yi, B., "Airway commensal bacteria in cystic fibrosis inhibit the growth of *P. aeruginosa* via a released metabolite," *Microbiological Research* 2024;283: 127680. https://doi.org/10.1016/j.micres.2024.127680

11. Banerjee, S., Schlaeppi, K., and van der Heijden, M.G.A., "Keystone taxa as drivers of microbiome structure and functioning," *Nature Reviews Microbiology* 2018;16: 567–76. https://doi.org/10.1038/s41579-018-0024-1

12. Kong, F., Hua, Y., Zeng, B., Ning, R., Li, Y., and Zhao, J., "Gut microbiota signatures of longevity," *Current Biology* 2016;26(18): PR832–33. https://doi.org/10.1016/j.cub.2016.08.015

13. David, L. A., Weil, A., Ryan, E. T., Calderwood, S. B., Harris, J. B., Chowdhury, F., Begum, Y., Qadri, F., LaRocque, R. C., and Turnbaugh, P. J., "Gut microbial succession follows acute secretory diarrhea in humans," *mBio* 2015;6(3). https://doi.org/10.1128/mBio.00381-15

14. Qin, N., Yang, F., Li, A., Prifti, E., Chen, Y., Shao, L., Guo, J., Le Chatelier, E., Yao, J., Wu, L., Zhou, J., Ni, S., Liu, L., Pons, N., Batto, J. M., Kennedy, S. P., Leonard, P., Yuan, C., Ding, W., Chen, Y., Hu, X., Zheng,

B., Qian, G., Xu, W., Ehrlich, S. D., Zheng, S., and Li, L., "Alterations of the human gut microbiome in liver cirrhosis," *Nature* 2014;513: 59–64. https://doi.org/10.1038/nature13568

15. Liu, C., Frank, D. N., Horch, M., Chau, S., Ir, D., Horch, E. A., Tretina, K., van Besien, K., Lozupone, C. A., and Nguyen, V. H., "Associations between acute gastrointestinal GvHD and the baseline gut microbiota of allogeneic hematopoietic stem cell transplant recipients and donors," *Bone Marrow Transplantation* 2017;52(12): 1643–50. https://doi.org/10.1038/bmt.2017.200

16. Parfrey, L. W., Walters, W. A., and Knight, R., "Microbial eukaryotes in the human microbiome: Ecology, evolution, and future directions," *Frontiers in Microbiology* 2011;2: 153. https://doi.org/10.3389/fmicb.2011.00153

17. Hrncir, T., "Gut microbiota dysbiosis: Triggers, consequences, diagnostic and therapeutic options," *Microorganisms* 2022;10(3): 578. https://doi.org/10.3390/microorganisms10030578

18. McGhee, J. D., "The *C. elegans* intestine," May 3, 2007. In: WormBook: The Online Review of *C. elegans* Biology [Internet], Pasadena (CA): Worm-Book; 2005–2018. https://www.ncbi.nlm.nih.gov/books/NBK19717/

19. Thapa, M., Kumari, A., Chin, C. Y., Choby, J. E., Jin, F., Bogati, B., Chopyk, D. M., Koduri, N., Pahnke, A., Elrod, E. J., Burd, E. M., Weiss, D. S., and Grakoui, A., "Translocation of gut commensal bacteria to the brain," *bioRxiv* 2023. https://doi.org/10.1101/2023.08.30.555630

20. Chia, L. W., Hornung, B.V.H., Aalvink, S., Schaap, P. J., De Vos, W. M., Knol, J., and Belzer, C., "Deciphering the trophic interaction between *Akkermansia muciniphila* and the butyrogenic gut commensal *Anaerostipes caccae* using a metatranscriptomic approach," *Antonie van Leeuwenhoek* 2018;111(6): 859–73. https://doi.org/10.1007/s10482-018-1040-x

21. Burger-van Paassen, N., Vincent, A., Puiman, P. J., van der Sluis, M., Bouma, J., Boehm, G., van Goudoever, J. B., van Seuningen, I., and Renes, I. B., "The regulation of intestinal mucin MUC2 expression by short-chain fatty acids: Implications for epithelial protection," *Biochemical Journal* 2009;420(2): 211–19. https://doi.org/10.1042/BJ20082222

22. Gaudier, E., Rival, M., Buisine, M. P., Robineau, I., and Hoebler, C., "Butyrate enemas upregulate *Muc* genes expression but decrease adherent mucus thickness in mice colon," *Physiological Research* 2009; 58(1): 111–19. https://doi.org/10.33549/physiolres.931271

23. Chen, J., and Vitetta, L., "The role of butyrate in attenuating pathobiont-induced hyperinflammation," *Immune Network* 2020;20(2): e15. https://doi.org/10.4110/in.2020.20.e15

24. Ferreira, T. M., Leonel, A. J., Melo, M. A., Santos, R. R., Cara, D. C.,

Cardoso, V. N., Correia, M. I., and Alvarez-Leite, J. I., "Oral supplementation of butyrate reduces mucositis and intestinal permeability associated with 5-fluorouracil administration," *Lipids* 2012;47(7): 669–78. https://doi.org/10.1007/s11745-012-3680-3

25. Desai, M., Seekatz, A. M., Koropatkin, N. M., Kamada, N., Hickey, C. A., Wolter, M., Pudlo, N. A., Kitamoto, S., Terrapon, N., Muller, A., Young, V. B., Henrissat, B., Wilmes, P., Stappenbeck, T. S., Núñez, G., and Martens, E. C., "A dietary fiber-deprived gut microbiota degrades the colonic mucus barrier and enhances pathogen susceptibility," *Cell* 2016;167(5): 1339–53. https://doi.org/10.1016/j.cell.2016.10.043

26. Kõiv, V., and Tenson, T., "Gluten-degrading bacteria: Availability and applications," *Applied Microbiology and Biotechnology* 2021;105(8): 3045–59. https://doi.org/10.1007/s00253-021-11263-5

27. Brandt, A., Kromm, F., Hernández-Arriaga, A., Sánchez, M. I., Bozkir, H. Ö., Staltner, R., Baumann, A., Camarinha-Silva, A., Heijtz, R. D., and Bergheim, I., "Cognitive alterations in old mice are associated with intestinal barrier dysfunction and induced toll-like receptor 2 and 4 signaling in different brain regions," *Cells* 2023;12(17): 2153. https://doi.org/10.3390/cells12172153

28. Fullerton, J. N., Segre, E., De Maeyer, R. P., Maini, A. A., and Gilroy, D. W., "Intravenous endotoxin challenge in healthy humans: An experimental platform to investigate and modulate systemic inflammation," *JoVE* 2016(111): 53913. https://doi.org/10.3791/53913

29. Pinti, M., Cevenini, E., Nasi, M., De Biasi, S., Salvioli, S., Monti, D., Benatti, S., Gibellini, L., Cotichini, R., Stazi, M. A., Trenti, T., Franceschi, C., and Cossarizza, A., "Circulating mitochondrial DNA increases with age and is a familiar trait: Implications for 'inflamm-aging,'" *European Journal of Immunology* 2014;44(5): 1552–62. https://doi.org/10.1002/eji.201343921

30. Wesemann, D. R., Portuguese, A. J., Meyers, R. M., Gallagher, M. P., Cluff-Jones, K., Magee, J. M., Panchakshari, R. A., Rodig, S. J., Kepler, T. B., Alt, F. W., "Microbial colonization influences early B-lineage development in the gut lamina propria," *Nature* 2013;501: 112–15. https://doi.org/10.1038/nature12496

31. Zheng, D., Liwinski, T., and Elinav, E., "Interaction between microbiota and immunity in health and disease," *Cell Research* 2020;30(6): 492–506. https://doi.org/10.1038/s41422-020-0332-7

32. Hajam, I. A., Dar, P. A., Shahnawaz, I., Jaume, J. C., and Lee, J. H., "Bacterial flagellin—a potent immunomodulatory agent," *Experimental & Molecular Medicine* 2017;49(9): e373. https://doi.org/10.1038/emm.2017.172

33. Kavanagh, E., "Long Covid brain fog: A neuroinflammation phenomenon?" *Oxford Open Immunology* 2022;3(1): iqac007. https://doi.org/10.1093/oxfimm/iqac007

34. Tsounis, E. P., Triantos, C., Konstantakis, C., Marangos, M., and Assimakopoulos, S. F., "Intestinal barrier dysfunction as a key driver of severe COVID-19," *World Journal of Virology* 2023;12(2): 68–90. https://doi.org/10.5501/wjv.v12.i2.68

35. Bersano, A., Engele, J., and Schäfer, M.K.E., "Neuroinflammation and brain disease," *BMC Neurology* 2023;23: 227. https://doi.org/10.1186/s12883-023-03252-0

36. Thevaranjan, N., Puchta, A., Schulz, C., Naidoo, A., Szamosi, J. C., Verschoor, C. P., Loukov, D., Schenck, L. P., Jury, J., Foley, K. P., Schertzer, J. D., Larché, M. J., Davidson, D. J., Verdú, E. F., Surette, M. G., and Bowdish, D.M.E., "Age-associated microbial dysbiosis promotes intestinal permeability, systemic inflammation, and macrophage dysfunction," *Cell Host & Microbe* 2017;21(4): 455–66.e4. https://doi.org/10.1016/j.chom.2017.03.002

37. Thion, M. S., Ginhoux, F., and Garel, S., "Microglia and early brain development: An intimate journey," *Science* 2018;362(6411): 185–89. https://doi.org/10.1126/science.aat0474

38. Butler, C. A., Popescu, A. S., Kitchener, E.J.A., Allendorf, D. H., Puigdellívol, M., and Brown, G. C., "Microglial phagocytosis of neurons in neurodegeneration, and its regulation," *Journal of Neurochemistry* 2021;158(3): 621–39. https://doi.org/10.1111/jnc.15327

39. Roberts, R. C., Farmer, C. B., and Walker, C. K., "The human brain microbiome; there are bacteria in our brains!" *Psychiatry and Behavioral Neurobio*, Univ. of Alabama, Birmingham.

40. Fasano, A., "All disease begins in the (leaky) gut: Role of zonulin-mediated gut permeability in the pathogenesis of some chronic inflammatory diseases," *F1000Research* 2020;9(F1000 Faculty Rev): 69. https://doi.org/10.12688/f1000research.20510.1

CHAPTER 2: THE MANIPULATED BRAIN

1. Sumich, A., Heym, N., Lenzoni, S., and Hunter, K., "Gut microbiome-brain axis and inflammation in temperament, personality and psychopathology," *Current Opinion in Behavioral Sciences* 2022;44: 101101. https://doi.org/10.1016/j.cobeha.2022.101101

2. Tubbs, R. S., Rizk, E., Shoja, M. M., Loukas, M., Barbaro, N., and Spinner, R. J., *Nerves and Nerve Injuries: Vol 1: History, Embryology, Anatomy, Imaging, and Diagnostics* (Cambridge, Massachusetts: Academic Press), 2015.

3. Powell, N., Walker, M. M., and Talley, N. J., "The mucosal immune system: Master regulator of bidirectional gut-brain communications," *Nature Reviews Gastroenterology & Hepatology* 2017;14(3): 143–59. https://doi .org/10.1038/nrgastro.2016.191

4. Forsythe, P., Bienenstock, J., and Kunze, W. A., "Vagal pathways for microbiome-brain-gut axis communication," *Advances in Experimental Medicine and Biology* 2014;817: 115–33. https://doi.org/10.1007/978 -1-4939-0897-4_5

5. Braniste, V., Al-Asmakh, M., Kowal, C., Anuar, F., Abbaspour, A., Tóth, M., Korecka, A., Bakocevic, N., Guan, N. L., Kundu, P., Gulyás, B., Halldin, K., Hultenby, H., Nilsson, H., Hebert, H., Volpe, B. T., Diamond, B., and Pettersson, S., "The gut microbiota influences blood-brain barrier permeability in mice," *Science Translational Medicine* 2014;6(263): 263ra158. https://doi.org/10.1126/scitranslmed .3009759

6. Brekke, E., Morken, T. S., Walls, A. B., Waagepetersen, H., Schousboe, A., and Sonnewald, U., "Anaplerosis for glutamate synthesis in the neonate and in adulthood," *Advances in Neurobiology* 2016;13: 43–58. https://doi .org/10.1007/978-3-319-45096-4_3

7. Kaelberer, M. M., Buchanan, K. L., Klein, M. E., Barth, B. B., Montoya, M. M., Shen, X., and Bohórquez, D. V., "A gut-brain neural circuit for nutrient sensory transduction," *Science* 2018;361: eaat5236. https://doi.org/10.1126/science.aat5236

8. Meldrum, B. S., "Glutamate as a neurotransmitter in the brain: Review of physiology and pathology," *Journal of Nutrition* 2000;130(4): 1007S–15S. https://doi.org/10.1093/jn/130.4.1007S

9. Mitani, H., Shirayama, Y., Yamada, T., Maeda, K., Ashby, C. R., Jr., and Kawahara, R., "Correlation between plasma levels of glutamate, alanine and serine with severity of depression," *Progress in Neuro-Psychopharmacology and Biological Psychiatry* 2006;30(6): 1155–58. https://doi.org/10.1016/j .pnpbp.2006.03.036

10. Holemans, S., De Paermentier, F., Horton, R. W., Crompton, M. R., Katana, C.L.E., and Maloteaux, J., "NMDA glutamatergic receptors, labelled with [³H]MK-801, in brain samples from drug-free depressed suicides," *Brain Research* 1993;616(1–2): 138–43. https://doi .org/10.1016/0006-8993(93)90202-X

11. Frye, M. A., Tsai, G. E., Huggins, T., Coyle, J. T., and Post, R. M., "Low cerebrospinal fluid glutamate and glycine in refractory affective disorder," *Biological Psychiatry* 2007;61(2): 162–66. https://doi .org/10.1016/j.biopsych.2006.01.024

12. Picciotto, M. R., Higley, M. J., and Mineur, Y. S., "Acetylcholine as a neuromodulator: Cholinergic signaling shapes nervous system function and behavior," *Neuron* 2012;76(1): 116–29. https://doi.org/10.1016/j.neuron.2012.08.036.

13. Amenta, F., and Tayebati, S. K., "Pathways of acetylcholine synthesis, transport and release as targets for treatment of adult-onset cognitive dysfunction," *Current Medicinal Chemistry* 2008;15(5): 488–98. https://doi.org/10.2174/092986708783503203

14. Eicher, T. P., and Mohajeri, M. H., "Overlapping mechanisms of action of brain-active bacteria and bacterial metabolites in the pathogenesis of common brain diseases," *Nutrients* 2022;14(13): 2661. https://doi.org/10.3390/nu14132661

15. Picciotto, M. R., Higley, M. J., and Mineur, Y. S., "Acetylcholine as a neuromodulator: Cholinergic signaling shapes nervous system function and behavior," *Neuron* 2012;76(1): 116–29. https://doi.org/10.1016/j.neuron.2012.08.036

16. Ferreira-Vieira, T. H., Guimaraes, I. M., Silva, F. R., and Ribeiro, F. M., "Alzheimer's disease: Targeting the cholinergic system," *Current Neuropharmacology* 2016;14(1): 101–15. https://doi.org/10.2174/1570159X13666150716165726

17. Lee, S. E., Lee, Y., and Lee, G. H., "The regulation of glutamic acid decarboxylases in GABA neurotransmission in the brain," *Archives of Pharmacal Research* 2019;42: 1031–39. https://doi.org/10.1007/s12272-019-01196-z

18. Frost, G., Sleeth, M. L., Sahuri-Arisoylu, M., Lizarbe, B., Cerdan, S., Brody, L., Anastasovska, J., Ghourab, S., Hankir, M., Zhang, S., Carling, D., Swann, J. R., Gibson, G., Viardot, A., Morrison, D., Thomas, E. L., and Bell, J. D., "The short-chain fatty acid acetate reduces appetite via a central homeostatic mechanism," *Nature Communications* 2014;5: 3611. https://doi.org/10.1038/ncomms4611

19. Lydiard, R. B., "The role of GABA in anxiety disorders," *Journal of Clinical Psychiatry* 2003;64: 21–27. https://pubmed.ncbi.nlm.nih.gov/12662130/

20. Allen, M. J., Sabir, S., and Sharma, S., "GABA receptor," updated February 13, 2023. In: StatPearls [Internet], Treasure Island (FL): StatPearls Publishing; January 2024. https://www.ncbi.nlm.nih.gov/books/NBK526124/

21. Eisenhofer, G., Aneman, A., Friberg, P., Hooper, D., Fandriks, L., Lonroth, H., Hunyady, B., and Mezey, E., "Substantial production of dopamine in the human gastrointestinal tract," *Journal of Clinical Endocrinology & Metabolism* 1997;82(11): 3864–71. https://doi.org/10.1210/jcem.82.11.4339

22. Gershon, M. D., "5-Hydroxytryptamine (serotonin) in the gastrointestinal tract," *Current Opinion in Endocrinology & Diabetes and Obesity* 2013;20(1): 14–21. https://doi.org/10.1097/MED.0b013e32835bc703

23. Abdel-Haq, R., Schlachetzki, J.C.M., Glass, C. K., and Mazmanian S. K., "Microbiome-microglia connections via the gut-brain axis," *Journal of Experimental Medicine* 2019;216(1): 41–59. https://doi.org/10.1084/jem.20180794

24. Helton, S. G., and Lohoff, F. W., "Serotonin pathway polymorphisms and the treatment of major depressive disorder and anxiety disorders," *Pharmacogenomics* 2015;16(5): 541–53. https://doi.org/10.2217/pgs.15.15

25. Pascual, M., Ibáñez, F., and Guerri, C., "Exosomes as mediators of neuron-glia communication in neuroinflammation," *Neural Regeneration Research* 2020;15(5): 796–801. https://doi.org/10.4103/1673-5374.268893

26. Schönfeld, P., Wojtczak, A. B., Geelen, M.J.H., Kunz, W., and Wojtczak L., "On the mechanism of the so-called uncoupling effect of medium- and short-chain fatty acids," *Biochimica et Biophysica Acta* 1988;936(3): 280–88. https://doi.org/10.1016/0005-2728(88)90003-5

27. Soret, R., Chevalier, J., De Coppet, P., Derkinderen, P., Segain, J. P., and Neunlist, M., "Short-chain fatty acids regulate the enteric neurons and control gastrointestinal motility in rats," *Gastroenterology* 2010;138(5): 1772–82. https://doi.org/10.1053/j.gastro.2010.01.053

28. Resende, W. R., Valvassori, S. S., Réus, G. Z., Gislaine, Z., Varela, R. B., Arent, C. O., Ribeiro, K. F., Bavaresco, D. V., Andersen, M. L., Zugno, A. I., and Quevado, J., "Effects of sodium butyrate in animal models of mania and depression: Implications as a new mood stabilizer," *Behavioural Pharmacology* 2013;24(7): 569–79. https://doi.org/10.1097/FBP.0b013e32836546fc

29. Valvassori, S. S., Wilson, R. R., Budni, J., Dal-Pont, G. C., Baveresco, D. V., Reus, G. Z., Carvalho, A. F., Goncalves, C. L., Furlanetto, C. B., Streck, E. L., and Quevedo, J., "Sodium butyrate, a histone deacety-lase inhibitor, reverses behavioral and mitochondrial alterations in animal models of depression induced by early- or late-life stress," *Current Neurovascular Research* 2015;12(4): 312–20. https://doi.org/10.2174/1567202612666150728121121

30. Hoogland, I.C.M., Houbolt, C., van Westerloo, D.J., van Gool, W.A., and van de Beek, D., "Systemic inflammation and microglial activation: Systematic review of animal experiments," *Journal of Neuroinflammation* 2015;12: 114. https://doi.org/10.1186/s12974-015-0332-6

31. Erny, D., Hrabĕ de Angelis, A. L., Jaitin, D., Wieghofer, P., Staszewski, O., David, E., Keren-Shaul, H., Mahlakoiv, T., Jakobshagen, K., Buch, T.,

Schwierzeck, V., Utermöhlen, O., Chun, E., Garrett, W. S., McCoy, K. D., Diefenbach, A., Staeheli, P., Stecher, B., Amit, I., and Prinz, M., "Host microbiota constantly control maturation and function of microglia in the CNS," *Nature Neuroscience* 2015;18(7): 965–77. https://doi.org/10.1038/nn.4030

32. Soret, R., Chevalier, J., De Coppet, P., Poupeau, G., Derkinderen, P., Segain, J. P., and Neunlist, M., "Short-chain fatty acids regulate the enteric neurons and control gastrointestinal motility in rats," *Gastroenterology* 2010;138(5): 1772–82. https://doi.org/10.1053/j.gastro.2010.01.053

33. Liu, H., Wang, J., He, T., Becker, S., Zhang, G., Li, D., and Ma, X., "Butyrate: A double-edged sword for health?" *Advances in Nutrition* 2018;9(1): 21–29. https://doi.org/10.1093/advances/nmx009

34. Resende, W. R., Valvassori, S. S., Réus, G. Z., Varela, R. B., Arent, C. O., Ribeiro, K. F., Bavaresco, D. V., Andersen, M. L., Zugno, A. I., and Quevedo, J., "Effects of sodium butyrate in animal models of mania and depression: Implications as a new mood stabilizer," *Behavioural Pharmacology* 2013;24(7): 569–79. https://doi.org/10.1097/FBP.0b013e32836546fc

35. Valvassori, S. S., Wilson, R. R., Budni, J., Dal-Pont, G. C., Baveresco, D. V., Reus, G. Z., Carvalho, A. F., Goncalves, C. L., Furlanetto, C. B., Streck, E. L., and Quevedo, J., "Sodium butyrate, a histone deacetylase inhibitor, reverses behavioral and mitochondrial alterations in animal models of depression induced by early- or late-life stress," *Current Neurovascular Research* 2015;12(4): 312–20. https://doi.org/10.2174/1567202612666150728121121

36. Ahmed, H., Leyrolle, Q., Koistinen, V., Kärkkäinen, O., Layé, S., Delzenne, N., and Hanhineva, K., "Microbiota-derived metabolites as drivers of gut–brain communication," *Gut Microbes* 2022;14(1): 2102878. https://doi.org/10.1080/19490976.2022.2102878

37. Rothhammer, V., Mascanfroni, I. D., Bunse, L., Takenaka, M. C., Kenison, J. E., Mayo, L., Chao, C., Patel, B., Yan, R., Blain, M., Alvarez, J., Kébir, H., Anandasabapathy, N., Izquierdo, G., Jung, S., Obholzer, N., Pochet, N., Clish, C. B., Prinz, M., Prat, A., Antel, J., and Quintana, F. J., "Type I interferons and microbial metabolites of tryptophan modulate astrocyte activity and central nervous system inflammation via the aryl hydrocarbon receptor," *Nature Medicine* 2016;22(6): 586–97. https://doi.org/10.1038/nm.4106

38. Wei, G. Z., Martin, K. A., Xing, P. Y., Agrawal, R., Whiley, L., Wood, T. K., Hejndorf, S., Ng, Y. Z., Low, J.Z.Y., Rossant, J., Nechanitzky, R., Holmes, E., Nicholson, J. K., Tan, E., Matthews, P. M., and Petterson,

S., "Tryptophan-metabolizing gut microbes regulate adult neurogenesis via the aryl hydrocarbon receptor," *Proceedings of the National Academy of Sciences USA* 2012;118(27): e2021091118. https://doi.org/10.1073/pnas.2021091118

39. Jaglin, M., Rhimi, M., Philippe, C., Pons, N., Bruneau, A., Goustard, B., Daugé, V., Maguin, E., Naudon, L., and Rabot, S., "Indole, a signaling molecule produced by the gut microbiota, negatively impacts emotional behaviors in rats," *Frontiers in Neuroscience* 2018;12: 216. https://doi.org/10.3389/fnins.2018.00216

40. Pocivavsek, A., Wu, H.-Q., Potter, M. C., Elmer, G. I., Pellicciari, R., and Schwarcz, R., "Fluctuations in endogenous kynurenic acid control hippocampal glutamate and memory," *Neuropsychopharmacology* 2011;36(11): 2357–67. https://doi.org/10.1038/npp.2011.127

41. Rothhammer, V., Borucki, D. M., Tjon, E. C., Takenaka, M. C., Chao, C. C., Ardura-Fabregat, A., de Lima, K. A., Gutiérrez-Vázquez, C., Hewson, P., Staszewski, O., Blain, M., Healy, L., Neziraj, T., Borio, M., Wheeler, M., Dragin, L. L., Laplaud, D. A., Antel, J., Alvarez, J. I., Prinz, M., and Quintana, F. J., "Microglial control of astrocytes in response to microbial metabolites," *Nature* 2018;557(7707): 724–28. https://doi.org/10.1038/s41586-018-0119-x

42. Gao, K., Mu, C.-L., Farzi, A., and Zhu, W.-Y., "Tryptophan metabolism: A link between the gut microbiota and brain," *Advances in Nutrition* 2020;11(3): 709–23. https://doi.org/10.1093/advances/nmz127

43. Matsuzaki, R., Gunnigle, E., Geissen, V., Clarke, G., Nagpal, J., and Cryan, J. F., "Pesticide exposure and the microbiota-gut-brain axis," *ISME Journal* 2013;17(8): 1153–66. https://doi.org/10.1038/s41396-023-01450-9

44. Ye, Z.-H., Fan, D.-F., and Zhang, T.-Y., "A narrative review of methane in treating neurological diseases," *Medical Gas Research* 2023;13(4): 161–64. https://doi.org/10.4103/2045-9912.372663

45. Woller, S. A., Eddinger, K. A., Corr, M., and Yaksh, T. L., "An overview of pathways encoding nociception," *Clinical and Experimental Rheumatology* 2017;35(Suppl. 107, 5): 40–46. https://pubmed.ncbi.nlm.nih.gov/28967373/

46. Li, Y.-L., Wu, P.-F., Chen, J.-G., Wang, S., Han, Q.-Q., Li, D., Wang, W., Guan, X.-L., Li, D., Long, L. H., Huang, J. G., and Wang, F., "Activity-dependent sulfhydration signal controls N-methyl-D-aspartate subtype glutamate receptor-dependent synaptic plasticity *via* increasing d-serine availability," *Antioxidants & Redox Signaling* 2017;27(7): 398–414. https://doi.org/10.1089/ars.2016.6936

47. Kolluru, G. K., Shen, X., Bir, S. C., and Kevil, C. G., "Hydrogen sulfide

chemical biology: Pathophysiological roles and detection," *Nitric Oxide* 2013; 35: 5–20. https://doi.org/10.1016/j.niox.2013.07.002

48. Lu, Y.-R., Zhang, Y., Rao, Y.-B., and Hu, L.-W., "The changes in and relationship between plasma nitric oxide and corticotropin-releasing hormone in patients with major depressive disorder," *Clinical and Experimental Pharmacology and Physiology* 2018;45(1): 10–15. https://doi.org/10.1111/1440-1681.12826

49. Kolluru, G. K., Shen, X., Bir, S. C., and Kevil, C. G., "Hydrogen sulfide chemical biology: Pathophysiological roles and detection," *Nitric Oxide* 2013;35: 5–20. https://doi.org/10.1016/j.niox.2013.07.002

50. Sunico, C. R., Portillo, F., González-Forero, D., and Moreno-López, B., "Nitric oxide-directed synaptic remodeling in the adult mammal CNS," *Journal of Neuroscience* 2005;25(6): 1448–58. https://doi.org/10.1523/JNEUROSCI.4600-04.2005

51. Zhang, X. R., Wang, Y. X., Zhang, Z. J., Li, L., and Reynolds, G. P., "The effect of chronic antipsychotic drug on hypothalamic expression of neural nitric oxide synthase and dopamine D2 receptor in the male rat," *PLoS ONE* 2012;7(4): e33247. https://doi.org/10.1371/journal.ponc.0033247

52. Bschor, T., and Bauer, M., "Efficacy and mechanisms of action of lithium augmentation in refractory major depression," *Current Pharmaceutical Design* 2006;12(23): 2985–92. https://doi.org/10.2174/138161206777947650

53. Queiroga, C.S.F., Vercelli, A., and Vieira, H.L.A., "Carbon monoxide and the CNS: Challenges and achievements," *British Journal of Pharmacology* 2015;172(6): 1533–45. https://doi.org/10.1111/bph.12729

54. Trentini, J. F., O'Neill, J. T., Poluch, S., and Juliano, S. L., "Prenatal carbon monoxide impairs migration of interneurons into the cerebral cortex," *Neurotoxicology* 2016;53: 31–44. https://doi.org/10.1016/j.neuro.2015.11.002

55. Yinghao, Y., Nie, Q., Dong, L., Zhang, J., Liu, S. F., Song, W., Wang, X., Wu, G., and Song, D., "Hydrogen attenuates allergic inflammation by reversing energy metabolic pathway switch," *Scientific Reports* 2020;10: 1962. https://doi.org/10.1038/s41598-020-58999-0

56. Scheperjans, F., Avo, V., Pereira, P.A.B., Koskinen, K., Paulin, L., Pekkonen, E., Haapaniemi, E., Kaakkola, S., Eerola-Rautio, J., Pohja, M., Kinnunen, E., Murros, K., and Auvinen, P., "Gut microbiota are related to Parkinson's disease and clinical phenotype," *Movement Disorders* 2015;30(3): 350–58. https://doi.org/10.1002/mds.26069

57. Ostojic, S. M., "Inadequate production of H_2 by gut microbiota and Parkinson disease," *Trends in Endocrinology & Metabolism* 2018;29(5): P286–88. https://doi.org/10.1016/j.tem.2018.02.006

58. Altaany, Z., Alkaraki, A., Abu-siniyeh, A., Al Momani, W., and Taani, O., "Evaluation of antioxidant status and oxidative stress markers in thermal sulfurous springs residents," *Heliyon* 2019;5(11): e02885. https://doi.org/10.1016/j.heliyon.2019.e02885

59. Kajimura, M., Nakanishi, T., Takenouchi, T., Morikawa, T., Hishiki, T., Yukutake, Y., and Suematsu, M., "Gas biology: Tiny molecules controlling metabolic systems," *Respiratory Physiology & Neurobiology* 2012;184(2): 139–48. https://doi.org/10.1016/j.resp.2012.03.016

60. Wu, C., Zou, P., Feng, S., Zhu, L., Li, F., Cheng-Yi Liu, T., Duan, R., and Yang, L., "Molecular hydrogen: An emerging therapeutic medical gas for brain disorders," *Molecular Neurobiology* 2023;60: 1749–65. https://doi.org/10.1007/s12035-022-03175-w

61. Artamonov, M. Y., Martusevich, A. K., Pyatakovich, F. A., Minenko, I. A., Dlin, S. V., and LeBaron, T. W., "Molecular hydrogen: From molecular effects to stem cells management and tissue regeneration," *Antioxidants* 2023;12(3): 636. https://doi.org/10.3390/antiox12030636

62. Satoh, Y., "The potential of hydrogen for improving mental disorders," *Current Pharmaceutical Design* 2021;27(5): 695–702. https://doi.org/10.2174/1381612826666201113095938

63. Camandola, S., and Mattson, M. P., "Brain metabolism in health, aging, and neurodegeneration," *EMBO Journal* 2017;36(11): 1474–92. https://doi.org/10.15252/embj.201695810

64. Anderson, G., and Maes, M., "Gut dysbiosis dysregulates central and systemic homeostasis via suboptimal mitochondrial function: Assessment, treatment and classification implications," *Current Topics in Medicinal Chemistry* 2000;20(7): 524–39. https://doi.org/10.2174/1568026620666200131094445

65. Walker, M. A., Volpi, S., Sims, K. B., Walter, J. E., and Traggiai, E., "Powering the immune system: Mitochondria in immune function and deficiency," *Journal of Immunology Research* 2014;21: 164309. https://doi.org/10.1155/2014/164309

66. Ma, J., Coarfa, C., Qin, X., Bonnen, P. E., Milosavljevic, A., Versalovic, J., and Aagaard, K., "mtDNA haplogroup and single nucleotide polymorphisms structure human microbiome communities," *BMC Genomics* 2014;15: 257. https://doi.org/10.1186/1471-2164-15-257

67. Samczuk, P., Hady, H. R., Adamska-Patruno, E., Citko, A., Dadan, J., Barbas, C., Kretowski, A., and Ciborowski, M., "In-and-out molecular

changes linked to the type 2 diabetes remission after bariatric surgery: An influence of gut microbes on mitochondria metabolism," *International Journal of Molecular Sciences* 2018;19(12): 3744. https://doi.org/10.3390 /ijms19123744

68. Endres, K., and Friedland, K., "Talk to me—interplay between mitochondria and microbiota in aging," *International Journal of Molecular Sciences* 2023;24(13): 10818. https://doi.org/10.3390/ijms241310818.

69. *Cellular and Molecular Biology* 2015;61(4): 121–24.

70. Zhao, C., Deng, W., and Gage, F. H., "Mechanisms and functional implications of adult neurogenesis," *Cell* 2008;132(4): 645–60. https://doi .org/10.1016/j.cell.2008.01.033

71. Ribeiro, M. F., Santos, A. A., Afonso, M. B., Rodrigues, P. M., Sá Santos, S., Castro, R. E., Rodrigues, C.M.P., and Solá, S., "Diet-dependent gut microbiota impacts on adult neurogenesis through mitochondrial stress modulation," *Brain Communications* 2020;2(2): fcaa165. https://doi .org/10.1093/braincomms/fcaa165

72. El Hayek, L., Khalifeh, M., Zibara, V., Abi Assaad, R., Emmanuel, N., Karnib, N., El-Ghandour, R., Nasrallah, P., Bilen, M., Ibrahim, P., Younes, J., Abou Haidar, E., Barmo, N., Jabre, V., Stephan, J. S., and Sleiman, S. F., "Lactate mediates the effects of exercise on learning and memory through SIRT1-dependent activation of hippocampal brain-derived neurotrophic factor (BDNF)," *Journal of Neuroscience* 2019;39(13): 2369–82. https:// doi.org/10.1523/JNEUROSCI.1661-18.2019

73. Ni Lochlainn, M., Bowyer, R.C.E., Moll, J. M., García, M. P., Wadge, S., Baleanu, A. F., Nessa, A., Sheedy, A., Akdag, G., Hart, D., Raffaele, G., Seed, P. T., Murphy, C., Harridge, S.D.R., Welch, A. A., Greig, C., Whelan, K., and Steves, C. J., "Effect of gut microbiome modulation on muscle function and cognition: The PROMOTe randomised controlled trial," *Nature Communications* 2024;15(1): 1859. https://doi .org/10.1038/s41467-024-46116-y

74. Arrè, V., Mastrogiacomo, R., Balestra, F., Serino, G., Viti, F., Rizzi, F., Curri, M. L., Giannelli, G., Depalo, N., and Scavo, M. P., "Unveiling the potential of extracellular vesicles as biomarkers and therapeutic nanotools for gastrointestinal diseases," *Pharmaceutics* 2024;16(4): 567. https://doi .org/10.3390/pharmaceutics16040567

75. Donoso-Meneses, D., Figueroa-Valdés, A. I., Khoury, M., and Alcayaga-Miranda, F., "Oral administration as a potential alternative for the delivery of small extracellular vesicles," *Pharmaceutics* 2023;15(3): 716. https:// doi.org/10.3390/pharmaceutics15030716

76. Stentz, R., Carvalho, A. L., Jones, E. J., and Carding, S. R., "Fantastic

voyage: The journey of intestinal microbiota-derived microvesicles through the body," *Biochemical Society Transactions* 2018;46(5): 1021–27. https://doi.org/10.1042/BST20180114

77. Kang, C.-S., Ban, M., Choi, E.-J., Moon, H.-G., Jeon, J.-S., Kim, D.-K., Park, S.-K., Jeon, S. G., Roh, T.-Y., Myung, S.-J., Gho, Y. S., Kim, J. G., and Kim, Y.-K., "Extracellular vesicles derived from gut microbiota, especially *Akkermansia muciniphila*, protect the progression of dextran sulfate sodium-induced colitis," *PLoS ONE* 2013;8(10): e76520. https://doi.org/10.1371/journal.pone.0076520

78. Yaghoubfar, R., Behrouzi, A., Ashrafian, F., Shahryari, A., Moradi, H. R., Choopani, S., Hadifar, S., Vaziri, F., Nojoumi, S. A., Fateh, A., Khatami, S., and Siadat, S. D., "Modulation of serotonin signaling/metabolism by *Akkermansia muciniphila* and its extracellular vesicles through the gut-brain axis in mice," *Scientific Reports* 2020;10: 22119. https://doi.org/10.1038/s41598-020-79171-8

79. Choi, J., Kim, Y.-K., and Han, P.-L., "Extracellular vesicles derived from *Lactobacillus plantarum* increase BDNF expression in cultured hippocampal neurons and produce antidepressant-like effects in mice," *Experimental Neurobiology* 2019;28(2): 158–71. https://doi.org/10.5607/en.2019.28.2.158

80. Sun, D., Chen, P., Xi, Y., and Sheng, J., "From trash to treasure: The role of bacterial extracellular vesicles in gut health and disease," *Frontiers in Immunology* 2023;14: 1274295. https://doi.org/10.3389/fimmu.2023.1274295

81. Prangishvili, D., Holz, I., Stieger, E., Nickell, S., Kristjansson, J. K., and Zillig, W., "Sulfolobicins, specific proteinaceous toxins produced by strains of the extremely thermophilic archaeal genus *Sulfolobus*," *Journal of Bacteriology* 2000;182(10): 2985–88. https://doi.org/10.1128/JB.182.10.2985-2988.2000

82. Wang, Y., Hoffmann, J. P., Chou, C.-W., Höner zu Bentrup, K., Fuselier, J. A., Bitoun, J. P., Wimley, W. C., and Morici, L. A., "*Burkholderia thailandensis* outer membrane vesicles exert antimicrobial activity against drug-resistant and competitor microbial species," *Journal of Microbiology* 2020;58(7): 550–62. https://doi.org/10.1007/s12275-020-0028-1

83. Kadurugamuwa, J. L., and Beveridge, T. J., "Bacteriolytic effect of membrane vesicles from Pseudomonas aeruginosa on other bacteria including pathogens: Conceptually new antibiotics," *Journal of Bacteriology* 1996;178(10): 2767–74. https://doi.org/10.1128/jb.178.10.2767-2774.1996

84. Naviaux, R. K., "*Perspective*: Cell danger response biology—the new science that connects environmental health with mitochondria and the rising tide of chronic illness," *Mitochondrion* 2020;51: 40–45. https://doi.org/10.1016/j.mito.2019.12.005

85. Naviaux, R. K., "Antipurinergic therapy for autism—an in-depth review," *Mitochondrion* 2018;43: 1–15. https://doi.org/10.1016/j.mito .2017.12.007

86. Motori, E., Puyal, J., Toni, N., Ghanem, A., Angeloni, C., Malaguti, M., Cantelli-Forti, G., Berninger, B., Conzelmann, K.-K., Götz, M., Winkl-hofer, K. F., Hrelia, S., and Bergami, M., "Inflammation-induced alteration of astrocyte mitochondrial dynamics requires autophagy for mitochon-drial network maintenance," *Cell Metabolism* 2013;18(6): 844–59. https:// doi.org/10.1016/j.cmet.2013.11.005

87. Naviaux, R. K., "*Perspective*: Cell danger response biology—the new science that connects environmental health with mitochondria and the rising tide of chronic illness," *Mitochondrion* 2020;51: 40–45. https://doi .org/10.1016/j.mito.2019.12.005

88. Coderch, L., López, O., de la Maza, A., and Parra, J. L., "Ceramides and skin function," *American Journal of Clinical Dermatology* 2003;4(2): 107–29. https://doi.org/10.2165/00128071-200304020-00004

89. Kogot-Levin, A., and Saada, A., "Ceramide and the mitochondrial respiratory chain," *Biochimie* 2014;100: 88–94. https://doi.org/10.1016 /j.biochi.2013.07.027

90. Schwarz, E., Prabakaran, S., Whitfield, P., Major, H., Leweke, F. M., Koethe, D., McKenna, P., and Bahn, S., "High throughput lipidomic profiling of schizophrenia and bipolar disorder brain tissue reveals al-terations of free fatty acids, phosphatidylcholines, and ceramides," *Journal of Proteome Research* 2008;7(10): 4266–77. https://doi.org/10 .1021/pr800188y

91. Sommansson, A., Nylander, O., and Sjöblom, M., "Melatonin decreases duodenal epithelial paracellular permeability via a nicotinic receptor–dependent pathway in rats in vivo," *Journal of Pineal Research* 2013;54(3): 282–91. https://doi.org/10.1111/jpi.12013

92. Borisenkov, M. F., Dorogina, O. I., Popov, S. V., Smirnov, V. V., Pecherkina, A. A., and Symaniuk, E. E., "The positive association between melatonin-containing food consumption and older adult life satisfaction, psychoemo-tional state, and cognitive function," *Nutrients* 2024;16(7): 1064. https:// doi.org/10.3390/nu16071064

93. Naviaux, R. K., "Mitochondrial and metabolic features of salugenesis and the healing cycle," *Mitochondrion* 2023;70: 131–63. https://doi.org /10.1016/j.mito.2023.04.003

94. Naviaux, R. K., "Metabolic features and regulation of the healing cycle—a new model for chronic disease pathogenesis and treatment," *Mitochondrion* 2019;46: 278–97. https://doi.org/10.1016/j.mito.2018.08.001

95. Naviaux, R. K., "Metabolic features of the cell danger response," *Mitochondrion* 2014;16: 7–17. https://doi.org/10.1016/j.mito.2013.08.006

96. Naviaux, R. K., "Incomplete healing as a cause of aging: The role of mitochondria and the cell danger response," *Biology* (Basel) 2019;8(2):27. https://doi.org/10.3390/biology8020027

97. Barbour, J. A., and Turner, N., "Mitochondrial stress signaling promotes cellular adaptations," *International Journal of Cell Biology* 2014: 156020. https://doi.org/10.1155/2014/156020

98. Naviaux, R. K., "Antipurinergic therapy for autism—an in-depth review," *Mitochondrion* 2018;43: 1–15. https://doi.org/10.1016/j.mito.2017.12.007

99. Nesci, S., Spagnoletta, A., and Oppedisano, F., "Inflammation, mitochondria and natural compounds together in the circle of trust," *International Journal of Molecular Sciences* 2023;24(7): 6106. https://doi.org/10.3390/ijms24076106

CHAPTER 3: YOUR HUNGER HAS BEEN HIJACKED

1. Everard, A., Belzer, C., Geurts, L., Ouwerkerk, J. P., Druart, C., Bindels, L. B., Guiot, Y., Derrien, M., Muccioli, G. G., Delzenne, N. M., de Vos, W. M., and Cani, P. D., "Cross-talk between *Akkermansia muciniphila* and intestinal epithelium controls diet-induced obesity," *Proceedings of the National Academy of Sciences USA* 2013;110(22): 9066–71. https://doi.org/10.1073/pnas.1219451110

2. Tsai, C.-Y., Lu, H.-C., Chou, Y.-H., Liu, P.-Y., Chen, H.-Y., Huang, M.-C., Lin, C.-H., and Tsai, C.-N., "Gut microbial signatures for glycemic responses of GLP-1 receptor agonists in type 2 diabetic patients: A pilot study," *Frontiers in Endocrinology* 2022;12: 814770. https://doi.org/10.3389/fendo.2021.814770

3. Alang, N., and Kelly, C. R., "Weight gain after fecal microbiota transplantation," *Open Forum Infectious Diseases* 2015;2(1): ofv004. https://doi.org/10.1093/ofid/ofv004

4. Bäckhed, F., Ding, H., Wang, T., Hooper, L. V., Koh, G. Y., Nagy, A., Semenkovich, C. F., and Gordon, J. I., "The gut microbiota as an environmental factor that regulates fat storage," *Proceedings of the National Academy of Sciences USA* 2004;101(44): 15718–23. https://doi.org/10.1073/pnas.0407076101

5. Salas-Venegas, V., Flores-Torres, R. P., Rodríguez-Cortés, Y. M., Rodríguez-Retana, D., Ramírez-Carreto, R. J., Concepción-Carrillo, L. E., Pérez-Flores, L. J., Alarcón-Aguilar, A., López-Díazguerrero, N. E., Gómez-González, B., Chavarría, A., and Konigsberg, M., "The obese brain:

Mechanisms of systemic and local inflammation, and interventions to reverse the cognitive deficit," *Frontiers in Integrative Neuroscience* 2022;16: 798995. https://doi.org/10.3389/fnint.2022.798995

6. Virtue, A. T., McCright, S. J., Wright, J. M., Jimenez, M. T., Mowel, W. K., Kotzin, J. J., Joannas, L., Basavappa, M. G., Spencer, S. P., Clark, M. L., Eisennagel, S. H., Williams, A., Levy, M., Manne, S., Henrickson, S. E., Wherry, E. J., Thaiss, C. A., Elinav, E., and Henao-Mejia, J., "The gut microbiota regulates white adipose tissue inflammation and obesity via family of microRNAs," *Science Translational Medicine* 2019;11(496): eaav1892. https://doi.org/10.1126/scitranslmed.aav1892

7. Cani, P. D., Amar, J., Iglesias, M. A., Poggi, M., Knauf, C., Bastelica, D., Neyrinck, A. M., Fava, F., Tuohy, K. M., Chabo, C., Waget, A., Delmée, E., Cousin, B., Sulpice, T., Chamontin, B., Ferrières, J., Tanti, J.-F., Gibson, G. R., Casteilla, L., Delzenne, N. M., Alessi, M. C., Burcelin, R., "Metabolic endotoxemia initiates obesity and insulin resistance," *Diabetes* 2007;56(7): 1761–72. https://doi.org/10.2337/db06-1491

8. Mehta, N. N., McGillicuddy, F. C., Anderson, P. D., Hinkle, C. C., Shah, R., Pruscino, L., Tabita-Martinez, J., Sellers, K. F., Rickels, M. R., and Reilly, M. P., "Experimental endotoxemia induces adipose inflammation and insulin resistance in humans," *Diabetes* 2010;59(1): 172–81. https://doi.org/10.2337/db09-0367

9. André, P., Laugerette, F., and Féart, C., "Metabolic endotoxemia: A potential underlying mechanism of the relationship between dietary fat intake and risk for cognitive impairments in humans?" *Nutrients* 2019;11(8): 1887. https://doi.org/10.3390/nu11081887

10. Saiyasit, N., Chunchai, T., Prus, D., Suparan, K., Pittayapong, P., Apaijai, N., Pratchayasakul, W., Sripetchwandee, J., Chattipakorn, N., and Chattipakorn, S. C., "Gut dysbiosis develops before metabolic disturbance and cognitive decline in high-fat diet-induced obese condition," *Nutrition* 2020;69: 110576. https://doi.org/10.1016/j.nut.2019.110576

11. Jia, X., Chen, Q., Wu, H., Liu, H., Jing, C., Gong, A., and Zhang, Y., "Exploring a novel therapeutic strategy: The interplay between gut microbiota and high-fat diet in the pathogenesis of metabolic disorders," *Frontiers in Nutrition* 2023;10: 1291853. https://doi.org/10.3389/fnut.2023.1291853

12. Caesar, R., Tremaroli, V., Kovatcheva-Datchary, P., Cani, P. D., and Bäckhed, F., "Crosstalk between gut microbiota and dietary lipids aggravates WAT inflammation through TLR signaling," *Cell Metabolism* 2015;22(4): 658–68. https://doi.org/10.1016/j.cmet.2015.07.026

13. Sun, J., Germain, A., Kaglan, G., Servant, F., Lelouvier, B., Federici, M., Fernandez-Real, J. M., Sala, D. T., Neagoe, R. M., Bouloumié, A., and Burcelin, R., "The visceral adipose tissue bacterial microbiota provides a signature of obesity based on inferred metagenomic functions," *International Journal of Obesity* 2023;47: 1008–22. https://doi.org/10.1038/s41366-023-01341-1

14. Amar, J., Burcelin, R., Ruidavets, J. B., Cani, P. D., Fauvel, J., Alessi, M. C., Chamontin, B., and Ferriéres, J., "Energy intake is associated with endotoxemia in apparently healthy men," *American Journal of Clinical Nutrition* 2008;87(5): 1219–23. https://doi.org/10.1093/ajcn/87.5.1219

15. Fei, N., and Zhao, L., "An opportunistic pathogen isolated from the gut of an obese human causes obesity in germfree mice," *ISME Journal* 2013;7(4): 880–84. https://doi.org/10.1038/ismej.2012.153

16. Yue, H., Qiu, B., Jia, M., Liu, W., Guo X.-F., Li, N., Xu, Z.-X., Du, F.-L., Xu, T., and Li, D., "Effects of α-linolenic acid intake on blood lipid profiles: A systematic review and meta-analysis of randomized controlled trials," *Critical Reviews in Food Science and Nutrition* 2021;61(17): 2894–910. https://doi.org/10.1080/10408398.2020.1790496

17. Kangwan, N., Pratchayasakul, W., Kongkaew, A., Pintha, K., Chattipakorn, N., and Chattipakorn, S. C. "Perilla seed oil alleviates gut dysbiosis, intestinal inflammation and metabolic disturbance in obese-insulin-resistant rats," *Nutrients* 2021;13(9): 3141. https://doi.org/10.3390/nu13093141

18. Newman, N. K., Zhang, Y., Padiadpu, J., Miranda, C. L., Magana, A. A., Wong, C. P., Hioki, K. A., Pederson, J. W., Li, Z., Gurung, M., Bruce, A. M., Brown, K., Bobe, G., Sharpton, T. J., Shulzhenko, N., Maier, C. S., Stevens, J. F., Gombart, A. F., and Morgun, A., "Reducing gut microbiome-driven adipose tissue inflammation alleviates metabolic syndrome," *Microbiome* 2023;11: 208. https://doi.org/10.1186/s40168-023-01637-4

19. Zhang, C., "The gut flora-centric theory based on the new medical hypothesis of 'hunger sensation comes from gut flora': A new model for understanding the etiology of chronic diseases in human beings," *Austin Internal Medicine* 2018;3(3): 1030. https://austinpublishinggroup.com/austin-internal-medicine/fulltext/aim-v3-id1030.php

20. Zhang, C., "The gut flora-centric theory based on the new medical hypothesis of 'hunger sensation comes from gut flora': A new model for understanding the etiology of chronic diseases in human beings," *Austin Internal Medicine* 2018;3(3): 1030. https://austinpublishinggroup.com/austin-internal-medicine/fulltext/aim-v3-id1030.php

21. Leitão-Gonçalves, R., Carvalho-Santos, Z., Francisco, A. P., Fioreze, G. T., Anjos, M., Baltazar, C., Elias, A. P., Itskov, P. M., Piper, M.D.W., and Ribeiro, C., "Commensal bacteria and essential amino acids control food choice behavior and reproduction," *PLoS Biology* 2017;15(4): e2000862. https://doi.org/10.1371/journal.pbio.2000862

22. Zhang, C., Gong, W., Li, Z., Gao, D., and Gao, Y., "Research progress of gut flora in improving human wellness," *Food Science and Human Wellness* 2019;8(2): 102–05. https://doi.org/10.1016/j.fshw.2019.03.007

23. Li, H., Zhang, L., Li, J., Wu, Q., Qian, L., He, J., Ni, Y., Kovatcheva-Datchary, P., Yuan, R., Liu, S., Shen, L., Zhang, M., Sheng, B., Li, P., Kang, K., Wu, L., Fang, Q., Long, X., Wang, X., Li, Y., Ye, Y., Ye, J., Bao, Y., Zhao, Y., Xu, G., Liu, X., Panagiotou, G., Xu, A., and Jia, W., "Resistant starch intake facilitates weight loss in humans by reshaping the gut microbiota," *Nature Metabolism* 2024;6: 578–97. https://doi.org/10.1038/s42255-024-00988-y

24. Chen, J., Xiao, Y., Li, D., Zhang, S., u, Z., Zhang, Q., and Bai, W., "New insights into the mechanisms of high-fat diet mediated gut microbiota in chronic diseases," *iMeta* 2023;2(1): e69. https://doi.org/10.1002/imt2.69

25. Kim, S., Cho, Y. S., Kim, H. M., Chung, O., Kim, H., Jho, S., Seomun, H., Kim, J., Bang, W. Y., Kim, C., An, J., Bae, C. H., Bhak, Y., Jeon, S., Yoon, H., Kim, Y., Jun, J., Lee, H., Cho, S., Uphyrkina, O., Kostyria, A., Goodrich, J., Miquelle, D., Roelke, M., Lewis, J., Yurchenko, A., Bankevich, A., Cho, J., Le, S., Edwards, J. S., Weber, J. A., Cook, J., Kim, S., Lee, H., Manica, A., Lee, I., O'brien, S. J., Bhak, J., and Yeo, J.-H., "Comparison of carnivore, omnivore, and herbivore mammalian genomes with a new leopard assembly," *Genome Biology* 2016;17: 211. https://doi.org/10.1186/s13059-016-1071-4

26. Zhang, C., "The gut flora-centric theory based on the new medical hypothesis of 'hunger sensation comes from gut flora': A new model for understanding the etiology of chronic diseases in human beings," *Austin Internal Medicine* 2018;3(3): 1030. https://austinpublishinggroup.com/austin-internal-medicine/fulltext/aim-v3-id1030.php

27. Pendyala, S., Walker, J. M., and Holt, P. R., "A high-fat diet is associated with endotoxemia that originates from the gut," *Gastroenterology* 2012;142(5): 1100–01.e2. https://doi.org/10.1053/j.gastro.2012.01.034

28. de La Serre, C. B., de Lartigue, G., and Raybould, H. E., "Chronic exposure to low dose bacterial lipopolysaccharide inhibits leptin signaling in vagal afferent neurons," *Physiology & Behavior* 2015;139: 188–94. https://doi.org/10.1016/j.physbeh.2014.10.032

29. Cho, Y. E., Kim, D.-K., Seo, W., Gao, B., Yoo, S.-H., and Song, B.-J., "Fructose promotes leaky gut, endotoxemia, and liver fibrosis through ethanol-inducible cytochrome P450-2E1–mediated oxidative and nitrative stress," *Hepatology* 2021;73(6): 2180–95. https://doi.org/10.1002/hep.30652

30. Guney, C., Bal, N. B., and Akar, F., "The impact of dietary fructose on gut permeability, microbiota, abdominal adiposity, insulin signaling and reproductive function," *Heliyon* 2023;9(8): e18896. https://doi.org/10.1016/j.heliyon.2023.e18896

31. Yan, J., Zheng, K., Zhang, X., and Jiang, Y., "Fructose consumption is associated with a higher risk of dementia and Alzheimer's disease: A prospective cohort study," *Journal of Prevention of Alzheimer's Disease* 2023;10(2):186–92. https://doi.org/10.14283/jpad.2023.7

32. Turnbaugh, P. J., Bäckhed, F., Fulton, L., and Gordon, J. I., "Diet-induced obesity is linked to marked but reversible alterations in the mouse distal gut microbiome," *Cell Host & Microbe* 2008;3(4): 213–23. https://doi.org/10.1016/j.chom.2008.02.015

33. Barnett, J. A., Bandy, M. L., and Gibson, D. L., "Is the use of glyphosate in modern agriculture resulting in increased neuropsychiatric conditions through modulation of the gut-brain-microbiome axis?" *Frontiers in Nutrition* 2022;9: 827384. https://doi.org/10.3389/fnut.2022.827384

34. Pappolla, M. A., Perry, G., Fang, X., Zagorski M., Sambamurti, K., and Poeggeler, B., "Indoles as essential mediators in the gut-brain axis: Their role in Alzheimer's disease," *Neurobiology of Disease* 2021;156: 105403. https://doi.org/10.1016/j.nbd.2021.105403

35. Zelante, T., Iannitti, R. G., Cunha, C., De Luca, A., Giovannini, G., Pieraccini, G., Zecchi, R., D'Angelo, C., Massi-Benedetti, C., Fallarino, F., Carvalho, A., Puccetti, P., and Romani, L., "Tryptophan catabolites from microbiota engage aryl hydrocarbon receptor and balance mucosal reactivity via interleukin-22," *Cell* 2013;39(2): 372–85. https://doi.org/10.1016/j.immuni.2013.08.003

36. Hill-Burns, E. M., Debelius, J. W., Morton, J. T., Wissemann, W. T., Lewis, M. R., Wallen Z. D., Peddada, S. D., Factor, S. A., Molho, E., Zabetian, C. P., Knight, R., and Payami, H., "Parkinson's disease and Parkinson's disease medications have distinct signatures of the gut microbiome," *Movement Disorders* 2017;32(5): 739–49. https://doi.org/10.1002/mds.26942

37. Nguyen, T. T., Hathaway, H., Kosciolek, T., Knight, R., and Jeste, D. V., "Gut microbiome in serious mental illnesses: A systematic review and

critical evaluation," *Schizophrenia Research* 2021;234: 24–40. https://doi
.org/10.1016/j.schres.2019.08.026

38. Chen, J., Zheng, P., Liu, Y. Y., Zhong, X. G., Wang, H. Y., Guo, Y. J., and
Xie, P., "Sex differences in gut microbiota in patients with major depressive disorder," *Neuropsychiatric Disease and Treatment* 2018;14: 647–55.
https://doi.org/10.2147/NDT.S159322

39. Sanacora, G., Mason, G. F., Rothman, D. L., Behar, K. L., Hyder,
F., Petroff, O.A.C., Berman, R. M., Charney, D. S., and Krystal, J. H.,
"Reduced cortical γ-aminobutyric acid levels in depressed patients
determined by proton magnetic resonance spectroscopy," *Archives of
General Psychiatry* 1999;56(11): 1043–47. https://doi.org/10.1001
/archpsyc.56.11.1043

40. Goddard, A. W., Mason, G. F., Almai, A., Rothman, D. L., Behar, K. L.,
Petroff, O. A., Charney, D. S., and Krystal, J. H., "Reductions in occipital
cortex GABA levels in panic disorder detected with 1H-magnetic resonance spectroscopy," *Archives of General Psychiatry* 2001;58(6): 556–61.
https://doi.org/10.1001/archpsyc.58.6.556

41. Epperson, C. N., Haga, K., Mason, G. F., Sellers, E., Gueorguieva, R.,
Zhang, W., Weiss, E., Rothman, D., and Krystal, J. H., "Cortical γ-
aminobutyric acid levels across the menstrual cycle in healthy women and
those with premenstrual dysphoric disorder: A proton magnetic resonance
spectroscopy study," *Archives of General Psychiatry* 2002;59(9): 851–58.
https://doi.org/10.1001/archpsyc.59.9.851

42. Barnett, J. A., and Gibson, D. L., "Separating the empirical wheat from
the pseudoscientific chaff: A critical review of the literature surrounding
glyphosate, dysbiosis and wheat-sensitivity," *Frontiers in Microbiology*
2020;11: 556729. https://doi.org/10.3389/fmicb.2020.556729

43. Christy, M.-L., Mamphweli, S., Meyer, E., and Okoh, A., "Antibiotic use
in agriculture and its consequential resistance in environmental sources:
Potential public health implications," *Molecules* 2018;23(4): 795. https://
doi.org/10.3390/molecules23040795

44. Kao-Ping, C., Fischer, M. A., and Linder, J. A., "Appropriateness of out-
patient antibiotic prescribing among privately insured US patients: ICD-
10-CM based cross sectional study," *BMJ* 2019;364: k5092. https://doi
.org/10.1136/bmj.k5092

45. Lu, K., Cable, P. H., Abo, R. P., Ru, H., Graffam, M. E., Schlieper,
K. A., Parry, N. M., Levine, S., Bodnar, W. M., Wishnok, J. S., Styblo,
M., Swenberg, J. A., Fox, J. G., and Tannenbaum, S. R., "Gut microbiome
perturbations induced by bacterial infection affect arsenic biotransforma-

tion," *Chemical Research in Toxicology* 2013;26(12): 1893–903. https://doi
.org/10.1021/tx4002868

46. Winston E. A., Wang, B., Sukhum, K. V., D'Souza, A. W., Hink, T.,
Cass, C., Seiler, S., Reske, K. A., Coon, C., Dubberke, E. R., Burnham,
C.A.D., Dantas, G., and Kwon, J. H., "Acute and persistent effects of
commonly used antibiotics on the gut microbiome and resistome in
healthy adults," *Cell* 2022;39(2): 110649. https://doi.org/10.1016/j
.celrep.2022.110649

47. Haak, B. W., Lankelma, J. M., Hugenholtz, F., Belzer, C., de Vos, W. M.,
and Wiersinga, W. J., "Long-term impact of oral vancomycin, cipro-
floxacin and metronidazole on the gut microbiota in healthy humans,"
Journal of Antimicrobial Chemotherapy 2019;74(3): 782–86. https://doi
.org/10.1093/jac/dky471

48. Dethlefsen, L., and Relman, D. A., "Incomplete recovery and indi-
vidualized responses of the human distal gut microbiota to repeated
antibiotic perturbation," *Proceedings of the National Academy of Sciences USA* 2010;108(Suppl. 1): 4554–61. https://doi.org/10.1073/pnas
.1000087107

49. Relman, D. A., "The human microbiome: Ecosystem resilience and
health," *Nutrition Reviews* 2012;70(Suppl. 1): S2–9. https://doi.org
/10.1111/j.1753-4887.2012.00489.x

50. Shan, J., Li, T., Zhou, X., Qin, W., Wang, Z., and Liao, Y., "Antibiotic drug piperacillin induces neuron cell death through
mitochondrial dysfunction and oxidative damage," *Canadian Journal of Physiology and Pharmacology* 2017;96(6): 562–68. https://doi.org/10.1139/cjpp
-2016-0679

51. Stefano, G. B., Samuel, J., and Kream, R., "Antibiotics may trigger mitochondrial dysfunction inducing psychiatric disorders," *Medical Science
Monitor: International Medical Journal of Experimental and Clinical Research* 2017;23: 101–06. https://doi.org/10.12659/MSM.899478

52. Liu, Z., Wei, S., Chen, X., Liu, L., Wei, Z., Liao, Z., Wu, J., Li, Z.,
Zhou, H., and Wang, D., "The effect of long-term or repeated use
of antibiotics in children and adolescents on cognitive impairment
in middle-aged and older person(s) adults: A cohort study," *Frontiers in Aging Neuroscience* 2022;14: 833365. https://doi.org/10.3389
/fnagi.2022.833365

53. Mehta, R. S., Lochhead, P., Wang, Y., Ma, W., Nguyen, L. H., Kochar, B.,
Huttenhower, C., Grodstein, F., and Chan, A. T., "Association of midlife
antibiotic use with subsequent cognitive function in women," *PLoS ONE*
2022;17(3): e0264649. https://doi.org/10.1371/journal.pone.0264649

54. Welu, J., Metzger, J., Bebensee, S., Ahrendt, A., and Vasek, M., "Proton pump inhibitor use and risk of dementia in the veteran population," *Federal Practitioner* 2019;36(Suppl. 4): S27–31. https://www.ncbi.nlm.nih.gov/pmc/articles/PMC6604981/

55. News from AAIC24: "Surprising differences found in how sleep medications increase dementia risk for some, protect others," AAIC press release, July 15, 2019. https://aaic.alz.org/releases_2019/monSLEEP-jul15.asp

56. Li, W., Jiang, J., Zhang, S., and Xiao, S., "Prospective association of general anesthesia with risk of cognitive decline in a Chinese elderly community population," *Scientific Reports* 2023;13: 13458. https://doi.org/10.1038/s41598-023-39300-5

57. Hampi, R., and Stárka, L., "Enocrine disrupters and gut microbiome interactions," *Physiological Research* 2020;69(Suppl. 2): S211–23. https://doi.org/10.33549/physiolres.934513

CHAPTER 4: THE DOSE MAKES THE POISON

1. Rougé, P., Culerrier, R., Granier, C., Rancé, F., and Barre, A., "Characterization of IgE-binding epitopes of peanut (*Arachis hypogaea*) PNA lectin allergen cross-reacting with other structurally related legume lectins," *Molecular Immunology* 2010;47(14): 2359–66. https://doi.org/10.1016/j.molimm.2010.05.006

2. Coelho, L. P., Kultima, J. R., Costea, P. I., Fournier, C., Pan, Y., Czarnecki-Maulden, G., Hayward, M. R., Forslund, S. K., Schmidt, T.S.B., Descombes, P., Jackson, J. R., Li, Q., and Bork, P., "Similarity of the dog and human gut microbiomes in gene content and response to diet," *Microbiome* 2018;6(72). https://doi.org/10.1186/s40168-018-0450-3

3. Ritter, S., "Monitoring and maintenance of brain glucose supply: Importance of hindbrain catecholamine neurons in this multifaceted task," in: Harris, R.B.S., ed., *Appetite and Food Intake: Central Control*, 2nd edition (Boca Raton, FL: CRC Press/Taylor & Francis, 2017), chapter 9. https://doi.org/10.1201/9781315120171-9

4. Braun-Fahrländer, C., Riedler, J., Herz, U., Eder, W., Waser, M., Grize, L., Maisch, S., Carr, D., Gerlach, F., Bufe, A., Lauener, R. P., Schierl, R., Renz, H., Nowak, D., and von Mutius, E. (Allergy and Endotoxin Study Team), "Environmental exposure to endotoxin and its relation to asthma in school-age children," *New England Journal of Medicine* 2002;347(12): 869–77. https://doi.org/10.1056/NEJMoa020057

5. Mizobuchi, H., and Soma, G.-I., "Low-dose lipopolysaccharide as an

immune regulator for homeostasis maintenance in the central nervous system through transformation to neuroprotective microglia," *Neural Regeneration Research* 2021;16(10): 1928–34. https://doi.org/10.4103/1673 -5374.308067

6. Neher, J. J., and Cunningham, C., "Priming microglia for innate immune memory in the brain," *Trends in Immunology* 2019;40(4): 358–74. https:// doi.org/10.1016/j.it.2019.02.001

7. Mizobuchi, H., "Oral route lipopolysaccharide as a potential dementia preventive agent inducing neuroprotective microglia," *Frontiers in Immunology* 2023;14: 1110583. https://doi.org/10.3389/fimmu.2023.1110583

8. Kobayashi Y., Inagawa, H., Kohchi, C., Kazumura, K., Tsuchiya, H., Miwa, T., Okazaki, K., and Soma, G.-I., "Oral administration of *Pantoea agglomerans*-derived lipopolysaccharide prevents metabolic dysfunction and Alzheimer's disease-related memory loss in senescence-accelerated prone 8 (SAMP8) mice fed a high-fat diet," *PLoS ONE* 2018;13(6): e0198493. https://doi.org/10.1371/journal.pone.0198493

9. Mizobuchi, H., "Oral route lipopolysaccharide as a potential dementia preventive agent inducing neuroprotective microglia," *Frontiers in Immunology* 2023;14: 1110583. https://doi.org/10.3389/fimmu.2023.1110583

10. Chen, Z., Lu, J., Srinivasan, N., Tan, B.K.H., and Chan, S. H., "Polysaccharide-protein complex from *Lycium barbarum* L. is a novel stimulus of dendritic cell immunogenicity," *Journal of Immunology* 2009;182(6): 3503–09. https://doi.org/10.4049/jimmunol.0802567

11. Shi, X., Wei, W., and Wang, N., "Tremella polysaccharides inhibit cellular apoptosis and autophagy induced by Pseudomonas aeruginosa lipopolysaccharide in A549 cells through sirtuin 1 activation," *Oncology Letters* 2018;15(6): 9609–16. https://doi.org/10.3892/ol.2018.8554

12. Chandrasekaran, K., Salimian, M., Konduru, S. R., Choi, J., Kumar, P., Long, A., Klimova, N., Ho, C.-Y., Kristian, T., and Russell, J. W., "Overexpression of Sirtuin 1 protein in neurons prevents and reverses experimental diabetic neuropathy," *Brain* 2019;142(12): 3737–52. https://doi .org/10.1093/brain/awz324

13. Seeley, J. J., and Ghosh, S., "Molecular mechanisms of innate memory and tolerance to LPS," *Journal of Leukocyte Biology* 2017;101(1): 107–19. https://doi.org/10.1189/jlb.3MR0316-118RR

14. Im, E., Riegler, F. M., Pothoulakis, C., and Rhee, S. H., "Elevated lipopolysaccharide in the colon evokes intestinal inflammation, aggravated in immune modulator-impaired mice," *American Journal of Physiology-Gastrointestinal and Liver Physiology* 2012;303(4): G490–97. https://doi .org/10.1152/ajpgi.00120.2012

15. Zhou, M., Zhang, H., Xu, X., Chen, H., and Qi, B., "Association between circulating cell-free mitochondrial DNA and inflammation factors in noninfectious diseases: A systematic review," *PLoS ONE* 2024;19(1): e0289338. https://doi.org/10.1371/journal.pone.0289338

16. Lei, T., Li, H., Fang, Z., Lin, J., Wang, S., Xiao, L., Yang, F., Liu, X., Zhang, J., Huang, Z., and Liao, W., "Polysaccharides from *Angelica sinensis* alleviate neuronal cell injury caused by oxidative stress," *Neural Regeneration Research* 2014;9(3): 260–67. https://doi.org/10.4103/1673-5374.128218

17. Nampoothiri, K. M., Beena, D. J., Vasanthakumari, D. S., and Ismail, B., Chapter 3: Health benefits of exopolysaccharides in fermented foods, in: Frias, J., Martinez-Villaluenga, C., and Peñas, E., eds., *Fermented Foods in Health and Disease Prevention* (Cambridge, Massachusetts: Academic Press, 2017): 49–62.

18. Nakata, H., Imamura, Y., Saha, S., Lobo, R. E., Kitahara, S., Araki, S., Tomokiyo, M., Namai, F., Hiramitsu, M., Inoue, T., Nishiyama, K., Villena, J., and Kitazawa, H., "Partial characterization and immunomodulatory effects of exopolysaccharides from *Streptococcus thermophilus* SBC8781 during soy milk and cow milk fermentation," *Foods* 2023;12(12): 2374. https://doi.org/10.3390/foods12122374

19. Bhandary, T., Kurian, C., Muthu, M., Anand, A., Anand, T., and Paari, K. A., "Exopolysaccharides derived from probiotic bacteria and their health benefits," *Journal of Pure and Applied Microbiology* 2023;17(1): 35–50. https://doi.org/10.22207/JPAM.17.1.40

20. Laiño. J., Villena, J., Kanmani, P., and Kitazawa, H., "Immunoregulatory effects triggered by lactic acid bacteria exopolysaccharides: New insights into molecular interactions with host cells," *Microorganisms* 2016;4(3): 27. https://doi.org/10.3390/microorganisms4030027

21. Lebeer, S., Claes, I. J., Verhoeven, T.L.A., Vanderleyden, J., and De Keersmaecker, S.C.J., "Exopolysaccharides of *Lactobacillus rhamnosus* GG form a protective shield against innate immune factors in the intestine," *Microbial Biotechnology* 2011;4(3): 368–74. https://doi.org/10.1111/j.1751-7915.2010.00199.x

22. Xu, X., Peng, Q., Zhang, Y., Tian, D., Zhang, P., Huang, Y., Ma, L., Dia, V. P., Qiao, Y., and Shi, B., "Antibacterial potential of a novel *Lactobacillus casei* strain isolated from Chinese northeast sauerkraut and the antibiofilm activity of its exopolysaccharides," *Food & Function* 2020;11(5): 4697–706. https://doi.org/10.1039/D0FO00905A

23. Ashrafian, F., Keshavarz Azizi Raftar, S., Shahryari, A., Behrouzi, A., Yaghoubfar, R., Lari, A., Moradi, H. R., Khatami, S., Omrani, M. D., Vaziri,

F., Masotti, A., and Siadat, S. D., "Comparative effects of alive and pasteurized *Akkermansia muciniphila* on normal diet-fed mice," *Scientific Reports* 2012;11: 17898. https://doi.org/10.1038/s41598-021-95738-5

24. López-García, E., Benítez-Cabello, A., Arenas-de Larriva, A. P., Gutierrez-Mariscal, F. M., Pérez-Martínez, P., Yubero-Serrano, E. M., Garrido-Fernández, A., and Arroyo-López, F. N., "Oral intake of *Lactiplantibacillus pentosus* LPG1 produces a beneficial regulation of gut microbiota in healthy persons: A randomised, placebo-controlled, single-blind trial," *Nutrients* 2023;15(8): 1931. https://doi.org/10.3390/nu15081931

25. Mazzio, E., Barnes, A., Badisa, R., Fierros-Romero, G., Williams, H., Council, S., and Soliman, K.F.A., "Functional immune boosters; the herb or its dead microbiome? Antigenic TLR4 agonist MAMPs found in 65 medicinal roots and algaes," *Journal of Functional Foods* 2023; 107: 105687. https://doi.org/10.1016/j.jff.2023.105687

26. Wei, X., Yang, W., Wang, J., Zhang, Y., Wang, Y., Long, Y., Tan, B., and Wan, X., "Health effects of whole grains: A bibliometric analysis," *Foods* 2022;11(24): 4094. https://doi.org/10.3390/foods11244094

27. De Punder, K., and Pruimboom, L., "The dietary intake of wheat and other cereal grains and their role in inflammation," *Nutrients* 2013;5 (3): 771–87. https://doi.org/10.3390/nu5030771

28. Plattner, V. E., Germann, B., Neuhaus, W., Noe, C. R., Gabor, F., and Wirth, M., "Characterization of two blood-brain barrier mimicking cell lines: Distribution of lectin-binding sites and perspectives for drug delivery," *International Journal of Pharmaceutics*, 2010;387(1–2): 34–41. https://doi.org/10.1016/j.ijpharm.2009.11.030

29. Taniguchi, Y., Yoshioka, N., Nishizawa, T., Inagawa, H., Kohchi, C., and Soma G.-I., "Utility and safety of LPS-based fermented flour extract as a macrophage activator," *Anticancer Research* 2009;29(3): 859–64. PMID: 19414320.

30. Kobayashi, Y., Inagawa, H., Kohchi, C., Kazumura, K., Tsuchiya, H., Okazaki, K., and Soma, G.-I., "Oral administration of *Pantoea agglomerans*-derived lipopolysaccharide prevents development of atherosclerosis in high-fat diet-fed apoE-deficient mice via ameliorating hyperlipidemia, pro-inflammatory mediators and oxidative responses," *PLoS ONE* 2018;13(3): e0195008. https://doi.org/10.1371/journal.pone.0195008

31. Dutkiewicz, J., Mackiewicz, B., Lemieszek, M. K., Golec, M., and Milanowski, J., "Pantoea agglomerans: A mysterious bacterium of evil and good. Part IV. Beneficial effects," *Annals of Agricultural and Environmental Medicine* 2016;23(2): 206–22. https://doi.org/10.5604/12321966.1203879

32. Montenegro, D., Kalpana, K., Chrissian, C., Sharma, A., Takaoka, A., Iacovidou, M., Soll, C. E., Aminova, O., Heguy, A., Cohen, L., Shen, S., and Kawamura, A., "Uncovering potential 'herbal probiotics' in Juzen-taiho-to through the study of associated bacterial populations," *Bioorganic & Medicinal Chemistry Letters* 2015;25(3): 466–69. https://doi.org/10.1016/j.bmcl.2014.12.036

33. Takaku, S., Shimizu, M., and Morita, R., "Japanese Kampo medicine juzentaihoto improves antiviral cellular immunity in tumour-bearing hosts," *Evidence-Based Complementary and Alternative Medicine* 2022: 6122955. https://doi.org/10.1155/2022/6122955

34. Ito, M., Maruyama, Y., Kitamura, K., Kobayashi, T., Takahashi, H., Yamanaka, N., Harabuchi, Y., Origasa, H., and Yoshizaki, T., "Randomized controlled trial of juzen-taiho-to in children with recurrent acute otitis media," *Auris Nasus Larynx* 2017;44(4): 390–97. https://doi.org/10.1016/j.anl.2016.10.002

35. Liu, H., Wang, J., and Tabira, T., "Juzen-taiho-to, an herbal medicine, promotes the differentiation of transplanted bone marrow cells into microglia in the mouse brain injected with fibrillar amyloid β," *Tohoku Journal of Experimental Medicine* 2014;233(2): 113–22. https://doi.org/10.1620/tjem.233.113

36. Elvira-Recuenco, M., and van Vuurde, J.W.L., "Natural incidence of endophytic bacteria in pea cultivars under field conditions," *Canadian Journal of Microbiology* 2000;46(11): 1036–41. https://doi.org/10.1139/w00-098

37. Patakova, P., Vasylkivska, M., Sedlar, K., Jureckova, K., Bezdicek, M., Lovecka, P., Branska, B., Kastanek, P., and Krofta, K., "Whole genome sequencing and characterization of *Pantoea agglomerans* DBM 3797, endophyte, isolated from fresh hop (*Humulus lupulus* L.)," *Frontiers in Microbiology* 2024;15: 1305338. https://doi.org/10.3389/fmicb.2024.1305338

38. Wastyk, H. C., Fragiadakis, G. K., Perelman, D., Dahan, D., Merrill, B. D., Yu, F. B., Topf, M., Gonzalez, C. G., Van Treuren, W., Han, S., Robinson, J. L., Elias, J. E., Sonnenburg, E. D., Gardner, C. D., and Sonnenburg, J. L., "Gut-microbiota-targeted diets modulate human immune status," *Cell* 2021;184(16): 4137–53.e14. https://doi.org/10.1016/j.cell.2021.06.019

CHAPTER 5: YOUR MICROBIOME PREDATES YOU

1. Kimura, I., Miyamoto, J., Ohue-Kitano, R., Watanabe, K., Yamada, T., Onuki, M., Aoki, R., Isobe, Y., Kashihara, D., Inoue, D., Inaba, A., Takamura, Y., Taira, S., Kumaki, S., Watanabe, M., Ito, M., Nakagawa, F., Irie, J., Kakuta,

H., Shinohara, M., Iwatsuki, K., Tsujimoto, G., Ohno, H., Arita, M., Itoh, H., and Hase, K., "Maternal gut microbiota in pregnancy influences off-spring metabolic phenotype in mice," *Science* 2020;367(6481): eaaw8429. https://doi.org/10.1126/science.aaw8429

2. Vuong, H. E., Pronovost, G. N., Williams, D. W., Coley, E.J.L., Siegler, E. L., Qiu A., Kazantsev, M., Wilson, C. J., Rendon, T., and Hsiao, E. Y., "The maternal microbiome modulates fetal neurodevelopment in mice," *Nature* 2020;586: 281–86. https://doi.org/10.1038/s41586-020-2745-3

3. Vuong, H. E., Pronovost, G. N., Williams, D. W., Coley, E.J.L., Siegler, E. L., Qiu A., Kazantsev, M., Wilson, C. J., Rendon, T., and Hsiao, E. Y., "The maternal microbiome modulates fetal neurodevelopment in mice," *Nature* 2020;586: 281–86. https://doi.org/10.1038/s41586-020-2745-3

4. Lavebratt, C., Yang, L. L., Giacobini, M., Forsell, Y., Schalling, M., Par-tonen, T., and Gissler, M., "Early exposure to antibiotic drugs and risk for psychiatric disorders: A population-based study," *Translational Psychiatry* 2019;9(317). https://doi.org/10.1038/s41398-019-0653-9

5. Lehikoinen, A. I., Kärkkäinen, O. K., Lehtonen, M.A.S., Auriola, S.O.K., Hanhineva, K. J., and Heinonen, S. T., "Alcohol and substance use are associated with altered metabolome in the first trimester serum sam-ples of pregnant mothers," *European Journal of Obstetrics & Gynecology and Reproductive Biology* 2018;223: 79–84. https://doi.org/10.1016/j.ejogrb.2018.02.004

6. Virdee, M. S., Saini, N., Kay, C. D., Neilson, A. P., Ting, S., Kwan, C., Hel-frich, K. K., Mooney, S. M., and Smith, S. M., "An enriched biosignature of gut microbiota-dependent metabolites characterizes maternal plasma in a mouse model of fetal alcohol spectrum disorder," *Scientific Reports* 2012;11: 248. https://doi.org/10.1038/s41598-020-80093-8

7. Sarker, G., and Peleg-Raibstein, D., "Maternal overnutrition induces long-term cognitive deficits across several generations," *Nutrients* 2019; 11(1): 7. https://doi.org/10.3390/nu11010007

8. Collado, M. C., Isolauri, E., Laitinen, K., and Salminen, S., "Effect of mother's weight on infant's microbiota acquisition, composition, and ac-tivity during early infancy: A prospective follow-up study initiated in early pregnancy," *American Journal of Clinical Nutrition* 2010;92(5): 1023–30. https://doi.org/10.3945/ajcn.2010.29877

9. Liu, X., Li, X., Xia, B., Jin, X., Zou, Q., Zeng, Z., Zhao, W., Yan, S., Li, L., Yuan, S., Zhao, S., Dai, X., Yin, F., Cadenas, E., Liu, R. H., Zhao, B., Hou, M., Liu, Z., and Liu, X., "High-fiber diet mitigates maternal obesity-induced cognitive and social dysfunction in the offspring via gut-brain

axis," *Cell Metabolism* 2021;33(5): 923–38.e6. https://doi.org/10.1016/j .cmet.2021.02.002

10. Rautava, S., Collado, M. C., Salminen, S., and Isolauri, E., "Probiotics modulate host-microbe interaction in the placenta and fetal gut: A randomized, double-blind, placebo-controlled trial," *Neonatology* 2012;102(3): 178–84. https://doi.org/10.1159/000339182

11. Dominguez-Bello, M. G., Costello, E. K., Contreras, M., Magris, M., Hidalgo, G., Fierer, N., and Knight, R., "Delivery mode shapes the acquisition and structure of the initial microbiota across multiple body habitats in newborns," *Proceedings of the National Academy of Sciences USA* 2010;107(26): 11971–75. https://doi.org/10.1073/pnas.1002601107

12. Jakobsson, H. E., Abrahamsson, T. R., Jenmalm, M. C., Harris, K., Quince, C., Jernberg, C., Björkstén, B., Engstrand, L., and Andersson, A. F., "Decreased gut microbiota diversity, delayed Bacteroidetes colonisation and reduced Th1 responses in infants delivered by Caesarean section," *Gut* 2014;63(4): 559–66. https://doi.org/10.1136/gutjnl-2012-303249

13. Biasucci, G., Rubini, M., Riboni, S., Morelli, L., Bessi, E., and Retetangos, C., "Mode of delivery affects the bacterial community in the newborn gut," *Early Human Development* 2010;86(Suppl. 1): 13–15. https.//doi .org/10.1016/j.earlhumdev.2010.01.004

14. Shao, Y., Forster, S. C., Tsaliki, E., Vervier, K., Strang, A., Simpson, N., Kumar, N., Stares, M. D., Rodger, A., Brocklehurst, P., Field, N., and Lawley, T. D., "Stunted microbiota and opportunistic pathogen colonization in caesarean-section birth," *Nature* 2019;574: 117–21. https://doi .org/10.1038/s41586-019-1560-1

15. Chu, D. M., Ma, J., Prince, A. L., Antony, K. M., Seferovic, M. D., and Aagaard, K. M., "Maturation of the infant microbiome community structure and function across multiple body sites and in relation to mode of delivery," *Nature Medicine* 2017; 23(3): 314–26. https://doi.org/10.1038 /nm.4272

16. Cabrera-Rubio, R., Collado, M. C., Laitinen, K., Salminen, S., Isolauri, E., and Mira, A., "The human milk microbiome changes over lactation and is shaped by maternal weight and mode of delivery," *American Journal of Clinical Nutrition* 2012;96(3): 544–51. https://doi.org/10.3945/ajcn.112.037382

17. Al Khalaf, S. Y., O'Neill, S. M., O'Keeffe, L. M., Henriksen, T. B., Kenny, L. C., Cryan, J. F., and Khashan, A. S., "The impact of obstetric mode of delivery on childhood behavior," *Social Psychiatry and Psychiatric Epidemiology* 2015;50: 1557–67. https://doi.org/10.1007/s00127 -015-1055-9

18. Curran, E. A., Kenny, L. C., Dalman, C., Kearney, P. M., Cryan, J. F., Dinan, T. G., and Khashan, A. S., "Birth by caesarean section and school performance in Swedish adolescents: A population-based study," *BMC Pregnancy and Childbirth* 2017;17(121). https://doi.org/10.1186/s12884-017-1304-x

19. Zhang, T., Sidorchuk, A., Sevilla-Cermeño, L., Vilaplana-Pérez, A., Chang, Z., Larsson, H., Mataix-Cols, D., and Fernández de la Cruz, L., "Association of cesarean delivery with risk of neurodevelopmental and psychiatric disorders in the offspring: A systematic review and meta-analysis," *JAMA Network Open* 2019;2(8): e1910236. https://doi.org/10.1001/jamanetworkopen.2019.10236

20. Xu, M., Yu, X., Fan, B., Li, G., and Ji, X., "Influence of mode of delivery on children's attention deficit hyperactivity disorder and childhood intelligence," *Psychiatry Investigation* 2023;20(8): 714–20. https://doi.org/10.30773/pi.2022.0310

21. Aarts, E., Ederveen, T.H.A., Naaijen, J., Zwiers, M. P., Boekhorst, J., Timmerman, H. M., Smeekens, S. P., Netea, M. G., Buitelaar, J. K., Franke, B., van Hijum, S.A.F.T., and Arias Vasquez, A., "Gut microbiome in ADHD and its relation to neural reward anticipation," *PLoS ONE* 2017;12(9): e0183509. https://doi.org/10.1371/journal.pone.0183509

22. Khanna, H. N., Roy, S., Shaikh, A., and Bandi, V., "Emerging role and place of probiotics in the management of pediatric neurodevelopmental disorders," *Euroasian Journal of Hepato-Gastroenterology* 2022;12(2):102–08. https://doi.org/10.5005/jp-journals-10018-1384

23. Castillo-Ruiz, A., Mosley, M., Jacobs, A. J., Hoffiz, Y. C., and Forger, N. G., "Birth delivery mode alters perinatal cell death in the mouse brain," *Proceedings of the National Academy of Sciences USA* 2018;115(46): 11826–31. https://doi.org/10.1073/pnas.1811962115

24. Morais, L. H., Golubeva, A. V., Moloney, G. M., Moya-Pérez, A., Ventura-Silva, A. P., Arboleya, S., Bastiaanssen, T. F., O'Sullivan, O., Rea, K., Borre, Y., Scott, K. A., Patterson, E., Cherry, P., Stilling, R., Hoban, A. E., El Aidy, S., Sequeira, A. M., Beers, S., Moloney, R. D., Renes, I. B., Wang, S., Knol, J., Paul Ross, R., O'Toole, P. W., Cotter, P. D., Stanton, C., Dinan, T. G., and Cryan, J. F., "Enduring behavioral effects induced by birth by caesarean section in the mouse," *Current Biology* 2020;30(19): 3761–74.e6. https://doi.org/10.1016/j.cub.2020.07.044

25. Robertson, R. C., Manges, A. R., Finlay, B. B., and Prendergast, A. J., "The human microbiome and child growth: First 1000 days and be-

yond," *Trends in Microbiology* 2019;27(2): 131–47. https://doi.org /10.1016/j.tim.2018.09.008

26. Sampson, T. R., and Mazmanian, S. K., "Control of brain development, function, and behavior by the microbiome," *Cell Host & Microbe* 2015;17(5): 565–76. https://doi.org/10.1016/j.chom.2015.04.011

27. Desbonnet, L., Clarke, G., Traplin, A., O'Sullivan, O., Crispie, F., Moloney, R. D., Cotter, P. D., Dinan, T. G., and Cryan, J. F., "Gut microbiota depletion from early adolescence in mice: Implications for brain and behaviour," *Brain, Behavior, and Immunity* 2015;48: 165–73. https://doi .org/10.1016/j.bbi.2015.04.004

28. Fox, M., Lee, S. M., Wiley, K. S., Lagishetty, V., Sandman, C. A., Jacops, J. P., and Glynn, L. M., "Development of the infant gut microbiome predicts temperament across the first year of life," *Development and Psychopathology* 2022;34(5): 1914–25. https://doi.org/10.1017/S0954579421000456

29. Lyons, K. E., Ryan, C. A., Dempsey, E. M., Ross, R. P., and Stanton, C., "Breast milk, a source of beneficial microbes and associated benefits for infant health," *Nutrients* 2020;12(4): 1039. https://doi.org/10.3390 /nu12041039

30. Ma, J., Li, Z., Zhang, W., Zhang, C., Zhang, Y., Mei, H., Zhuo, N., Wang, H., Wang, L., and Wu, D., "Comparison of gut microbiota in exclusively breast-fed and formula-fed babies: A study of 91 term infants," *Scientific Reports* 2020;10: 15792. https://doi.org/10.1038/s41598-020-72635-x

31. Soto, A., Martín, V., Jiménez, E., Mader, I., Rodríguez, J. M., and Fernández, L., "Lactobacilli and Bifidobacteria in human breast milk: Influence of antibiotherapy and other host and clinical factors," *Journal of Pediatric Gastroenterology and Nutrition* 2014;59(1): 78–88. https://doi .org/10.1097/MPG.0000000000000347

32. Stewart, C. J., Ajami, N. J., O'Brien, J. L., Hutchinson, D. S., Smith, D., Wong, M. C., Ross, M. C., Lloyd, R. E., Doddapaneni, H., Metcalf, G. A., Muzny, D., Gibbs, R. A., Vatanen, T., Huttenhower, C., Xavier, R. J., Rewers, M., Hagopian, W., Toppari, J., Ziegler, A.-G., She, J.-X., Akolkar, B., Lernmark, A., Hyoty, H., Vehik, K., Krischer, J. P., and Petrosino, J. F., "Temporal development of the gut microbiome in early childhood from the TEDDY study," *Nature* 2018;562: 583–88. https:// doi.org/10.1038/s41586-018-0617-x

33. Roger, L. C., Costabile, A., Holland, D. T., Hoyles, L., and McCartney, A. L., "Examination of faecal *Bifidobacterium* populations in breast- and formula-fed infants during the first 18 months of life," *Microbiology* 2010;156(11): 3329–41. https://doi.org/10.1099/mic.0.043224-0

34. Seferovic, M. D., Mohammad, M., Pace, R. M., Engevik, M., Versa-lovic, J., Bode, L., Haymond, M., and Aagaard, K. M., "Maternal diet alters human milk oligosaccharide composition with implications for the milk metagenome," *Scientific Reports* 2020;10: 22092. https://doi.org/10.1038/s41598-020-79022-6

35. Akbari, P., Fink-Gremmels, J., Willems, R.H.A.M., DiFilippo, E., Schols, H. A., Schoterman, M.H.C., Garssen, J., and Braber, S., "Characterizing microbiota-independent effects of oligosaccharides on intestinal epithelial cells: Insight into the role of structure and size: Structure-activity relationships of non-digestible oligosaccharides," *European Journal of Nutrition* 2017;56: 1919–30. https://doi.org/10.1007/s00394-016-1234-9

36. Hermansson, H., Kumar, H., Collado, M. C., Salminen, S., Isolauri, E., and Rautava, S., "Breast milk microbiota is shaped by mode of delivery and intrapartum antibiotic exposure," *Frontiers in Nutrition* 2019;6: 4. https://doi.org/10.3389/fnut.2019.00004

37. Demmer, E., Van Loon, M. D., Rivera, N., Rogers, T. S., Gertz, E. R., German, J. B., Smilowitz, J. T., and Zivkovic, A. M., "Addition of a dairy fraction rich in milk fat globule membrane to a high-saturated fat meal reduces the postprandial insulinaemic and inflammatory response in overweight and obese adults," *Journal of Nutritional Science* 2016;5: e14. https://www.ncbi.nlm.nih.gov/pmc/articles/PMC4791522/

38. Ji, X., Xu, W., Cui, J., Ma, Y., and Zhou, S., "Goat and buffalo milk fat globule membranes exhibit better effects at inducing apoptosis and reduc-tion the viability of HT-29 cells," *Scientific Reports* 2019;9: 2577. https://www.nature.com/articles/s41598-019-39546-y

39. Mohamed, H.J.J., Lee, E.K.H., Woo, K.C.K., Sarvananthan, R., Lee, Y. Y., and Mohd Hussin, Z. A., "Brain-immune-gut benefits with early life supplementation of milk fat globule membrane," *JGH Open* 2022;6(7): 454–61. https://doi.org/10.1002/jgh3.12775

40. Yuan, Q., Gong, H., Du, M., Li, T., and Mao, X., "Milk fat globule membrane supplementation to obese rats during pregnancy and lacta-tion promotes neurodevelopment in offspring *via* modulating gut mi-crobiota," *Frontiers in Nutrition* 2022;9: 945052. https://doi.org/10.3389/fnut.2022.945052

41. Timby, N., Domellöf, E., Hernell, O., Lönnerdal, B., and Domellöf, M., "Neurodevelopment, nutrition, and growth until 12 mo of age in infants fed a low-energy, low-protein formula supplemented with bo-vine milk fat globule membranes: A randomized controlled trial,"

American Journal of Clinical Nutrition 2014;99(4): 860–68. https://doi.org/10.3945/ajcn.113.064295

42. Schipper, L., Bartke, N., Marintcheva-Petrova, M., Schoen, S., Vandenplas, Y., and Hokken-Koelega, A.C.S., "Infant formula containing large, milk phospholipid-coated lipid droplets and dairy lipids affects cognitive performance at school age," *Frontiers in Nutrition* 2023;10: 1215199. https://doi.org/10.3389/fnut.2023.1215199

43. Timby, N., Domellöf, E., Hernell, O., Lönnerdal, B., and Domellöf, M., "Neurodevelopment, nutrition, and growth until 12 mo of age in infants fed a low-energy, low-protein formula supplemented with bovine milk fat globule membranes: A randomized controlled trial," *American Journal of Clinical Nutrition* 2014;99(4): 860–68. https://doi.org/10.3945/ajcn.113.064295

44. Su, Q., Wong, O.W.H., Lu, W., Wan, Y., Zhang, L., Xu, W., Li, M.K.T., Liu, C., Cheung, C. P., Ching, J.Y.L., Cheong, P. K., Leung, T. F., Chan, S., Leung, P., Chan, F.K.L., and Ng, S. C., "Multikingdom and functional gut microbiota markers for autism spectrum disorder," *Nature Microbiology* 2024;9: 2344–55. https://doi.org/10.1038/s41564-024-01739-1

45. Chen, Y., Xue, Y., Jia, L., Yang, M., Huang, G., and Xie, J., "Causal effects of gut microbiota on autism spectrum disorder: A two-sample mendelian randomization study," *Medicine* (Baltimore) 2024;103(9): e37284. https://doi.org/10.1097/MD.0000000000037284

46. Sharon, G., Cruz, N. J., Dae-Wook, K., Gandal, M. J., Wang, B., Kim, Y.-M., Zink, E. M., Casey, C. P., Taylor, B. C., Lane, C. J., Bramer, L. M., Isern, N. G., Hoyt, D. W., Noecker, C., Sweredoski, M. J., Moradian, A., Borenstein, E., Jansson, J. K., Knight, R., Metz, T. O., Lois, C., Geschwind, D. H., Krajmalnik-Brown, R., and Mazmanian, S. K., "Human gut microbiota from autism spectrum disorder promote behavioral symptoms in mice," *Cell* 2019;177(6): 1600–18.e17. https://doi.org/10.1016/j.cell.2019.05.004

47. Sharon, G., Cruz, N. J., Kang, D.-W., Gandal, M. J., Wang, B., Kim, Y.-M., Zink, E. M., Casey, C. P., Taylor, B. C., Lane, C. J., Bramer, L. M., Isern, N. G., Hoyt, D. W., Noecker, C., Sweredoski, M. J., Moradian, A., Borenstein, E., Jansson, J. K., Knight, R., Metz, T. O., Lois, C., Geschwind, D. H., Krajmalnik-Brown, R., and Mazmanian, S. K., "Human gut microbiota from autism spectrum disorder promote behavioral symptoms in mice," *Cell* 2019;177(6): 1600–18.e17. https://doi.org/10.1016/j.cell.2019.05.004

48. Siddiqui, M. F., Elwell, C., and Johnson, M. H., "Mitochondrial dysfunction

in autism spectrum disorders," *Autism Open Access* 2016;6(5): 1000190. https://doi.org/10.4172/2165-7890.1000190

49. Li, N., Chen, H., Cheng, Y., Xu, F., Ruan, G., Ying, S., Tang, W., Chen, L., Chen, M., Lv, L., Ping, Y., Chen, D., and Wei, Y., "Fecal microbiota transplantation relieves gastrointestinal and autism symptoms by improving the gut microbiota in an open-label study," *Frontiers in Cellular and Infection Microbiology* 2021;11: 759435. https://doi.org/10.3389/fcimb.2021.759435

50. Kang, D.-W., Adams, J. B., Coleman, D. M., Pollard, E. L., Maldonado, J., McDonough-Means, S., Caporaso, J. G., and Krajmalnik-Brown, R., "Long-term benefit of microbiota transfer therapy on autism symptoms and gut microbiota," *Scientific Reports* 2019;9: 5821. https://doi.org/10.1038/s41598-019-42183-0

51. Atladóttir, H. Ó., Thorsen, P., Østergaard, L., Schendel, D. E., Lemcke, S., Abdallah, M., and Parner, E. T., "Maternal infection requiring hospitalization during pregnancy and autism spectrum disorders," *Journal of Autism and Developmental Disorders* 2010;40: 1423–30. https://doi.org/10.1007/s10803-010-1006-y

52. Fowler, S. P., Gimeno Ruiz de Porras, D., Swartz, M. D., Stigler Granados, P., Heilbrun, L. P., and Palmer, R. F., "Daily early-life exposures to diet soda and aspartame are associated with autism in males: A case-control study," *Nutrients* 2023;15(17): 3772. https://doi.org/10.3390/nu15173772

53. Bian, X., Chi, L., Gao, B., Tu, P., Ru, H., and Lu, K., "Gut microbiome response to sucralose and its potential role in inducing liver inflammation in mice," *Frontiers in Physiology* 2017;8: 487. https://www.ncbi.nlm.nih.gov/pmc/articles/PMC5522834/

54. Jones, S. K., McCarthy, D. M., Stanwood, G. D., Schatschneider, C., and Bhide, P. G., "Learning and memory deficits produced by aspartame are heritable via the paternal lineage," *Scientific Reports* 2023;13: 14326. https://doi.org/10.1038/s41598-023-41213-2

55. Aagaard, K., Jepsen, J.R.M., Sevelsted, A., Horner, A., Vinding, R., Rosenberg, J. B., Brustad, N., Eliasen, A., Mohammadzadeh, P., Følsgaard, N., Hernández-Lorca, M., Fagerlund, B., Glenthøj, B. Y., Rasmussen, M. A., Bilenberg, N., Stokholm, J., Bønnelykke, K., Ebdrup, B. H., and Chawes, B., "High-dose vitamin D$_3$ supplementation in pregnancy and risk of neurodevelopmental disorders in the children at age 10: A randomized clinical trial," *American Journal of Clinical Nutrition* 2024;119(2): 362–70. https://doi.org/10.1016/j.ajcnut.2023.12.002

56. Jiang, Y., Dang, W., Sui, L., Gao, T., Kong, X., Guo, J., Liu, S., Nie, H., and

Jiang, Z., "Associations between vitamin D and core symptoms in ASD: An umbrella review," *Nutrition and Dietary Supplements* 2024;16: 59–91. https://doi.org/10.2147/NDS.S470462

57. Cantorna, M. T., Lin, Y.-D., Arora, J., Bora, S., Tian, Y., Nichols, R. G., and Patterson, A. D., "Vitamin D regulates the microbiota to control the numbers of RORγt/FoxP3+ regulatory T cells in the colon," *Frontiers in Immunology* 2019;10: 1772. https://doi.org/10.3389/fimmu.2019.01772

58. Zittermann, A., Ernst, J. B., Gummert, J. F,. and Borgermann, J., "Vitamin D supplementation, body weight and human serum 25-hydroxyvitamin D response: A systematic review," *European Journal of Nutrition* 2014;53(2): 367–74. https://doi.org/10.1007/s00394-013-0634-3

59. Cantarel, B. L., Waubant, E., Chehoud, C., Kuczynski, J., DeSantis, T. Z., Warrington, J., Venkatesan, A., Fraser, C. M., and Mowry, E. M., "Gut microbiota in multiple sclerosis: Possible influence of immunomodulators," *Journal of Investigative Medicine* 2015;63(5): 729–34. https://doi.org/10.1097/JIM.0000000000000192

60. Rose, S., Bennuri, S. C., Davis, J. E., Wynne, R., Slattery, J. C., Tippett, M., Delhey, L., Melnyk, S., Kahler, S. G., MacFabe, D. F., and Frye, R. E., "Butyrate enhances mitochondrial function during oxidative stress in cell lines from boys with autism," *Translational Psychiatry* 2018;8(42). https://doi.org/10.1038/s41398-017-0089-z

61. Miao, Z., Chen, L., Zhang, Y., Zhang, J., and Zhang, H., "*Bifidobacterium animalis* subsp. *lactis* Probio-M8 alleviates abnormal behavior and regulates gut microbiota in a mouse model suffering from autism," *mSystems* 2024;9(1): e01013–23. https://doi.org/10.1128/msystems.01013-23

CHAPTER 6: THE ADDICTION MICROBIOME

1. Alegado, R. A., and King, N., "Bacterial influences on animal origins," *Cold Spring Harbor Perspectives in Biology* 2014;6(11): a016162. https://doi.org/10.1101/cshperspect.a016162

2. Meade, E., and Garvey, M., "The role of neuro-immune interaction in chronic pain conditions; functional somatic syndrome, neurogenic inflammation, and peripheral neuropathy," *International Journal of Molecular Sciences* 2022;23(15): 8574. https://doi.org/10.3390/ijms23158574

3. Wang, G., Liu, Q., Guo, L., Zeng, H., Ding, C., Zhang, W., Xu, D., Wang, X., Qiu, J., Dong, Q., Fan, Z., Zhang, Q., and Pan, J., "Gut microbiota and relevant metabolites analysis in alcohol dependent

mice," *Frontiers in Microbiology* 2018;9: 1874. https://doi.org/10.3389 /fmicb.2018.01874

4. Bjørkhaug, S. T., Aanes, H., Neupane, S. P., Bramness, J. G., Malvik, S., Henriksen, C., Skar, V., Medhus, A. W., and Valeur, J., "Characterization of gut microbiota composition and functions in patients with chronic alcohol overconsumption," *Gut Microbes* 2019;10(6), 663–75. https://doi.org /10.1080/19490976.2019.1580097

5. Samuelson, D. R., Gu, M., Shellito, J. E., Molina, P. E., Taylor, C. M., Luo, M., and Welsh, D. A., "Intestinal microbial products from alcohol-fed mice contribute to intestinal permeability and peripheral immune activation," *Alcohol Clinical & Experimental Research* 2019;43(10): 2122–33. https://doi.org/10.1111/acer.14176

6. Peterson, V. L., Jury, N. J., Cabrera-Rubio, R., Draper, L. A., Crispie, F., Cotter, P. D., Dinan, T. G., Holmes, A., and Cryan, J. F., "Drunk bugs: Chronic vapour alcohol exposure induces marked changes in the gut microbiome in mice," *Behavioural Brain Research* 2017;323: 172–76. https:// doi.org/10.1016/j.bbr.2017.01.049.

7. Mutlu, E. A., Gillevet, P. M., Rangwala, H., Sikaroodi, M., Naqvi, A., Engen, P. A., Kwasny, M., Lau, C. K., and Keshavarzian, A., "Colonic microbiome is altered in alcoholism," *American Journal of Physiology-Gastrointestinal and Liver Physiology* 2012;302(9): G966–78. https://doi .org/10.1152/ajpgi.00380.2011

8. Blednov, Y. A., Ponomarev, I., Geil, C., Bergeson, S., Koob, G. F., and Harris, R. A., "Neuroimmune regulation of alcohol consumption: Behavioral validation of genes obtained from genomic studies," *Addiction Biology* 2012;17(1): 108–20. https://doi.org/10.1111/j.1369 -1600.2010.00284.x

9. Carbia, C., Bastiaanssen, T.F.S., Iannone, L. F., García-Cabrerizo, R., Boscaini, S., Berding, K., Strain, C. R., Clarke, G., Stanton, C., Dinan, T. G., and Cryan, J. F., "The microbiome-gut-brain axis regulates social cognition & craving in young binge drinkers," *eBioMedicine* 2023;89: 104442. https://doi.org/10.1016/j.ebiom.2023.104442

10. Yuan, Q., Yin, L., He, J., Zeng, Q., Liang, Y., Shen, Y., and Zu, X., "Metabolism of asparagine in the physiological state and cancer," *Cell Communication and Signaling* 2024;22(1). 163. https://doi.org/10.1186 /s12964-024-01540-x

11. Kärkkäinen, O., Klavus, A., Voutilainen, A., Virtanen, J., Lehtonen, M., Auriola, S., Kauhanen, J., and Rysä, J., "Changes in circulating metabolome precede alcohol-related diseases in middle-aged men: A prospective population-based study with a 30-year follow-up," *Alcohol Clinical &*

Experimental Research 2020;44(12): 2457–67. https://doi.org/10.1111/acer.14485

12. Leclercq, S., Matamoros, S., Cani, P. D., Neyrinck, A. M., Jamar, F., Stärkel, P., Windey, K., Tremaroli, V., Bäckhed, F., Verbeke, K., de Timary, P., and Delzenne, N. M., "Intestinal permeability, gut-bacterial dysbiosis, and behavioral markers of alcohol-dependence severity," *Proceedings of the National Academy of Sciences USA* 2014;111(42): E4485–93. https://doi.org/10.1073/pnas.1415174111

13. Bajaj, J. S., "Alcohol, liver disease and the gut microbiota," *Nature Reviews Gastroenterology & Hepatology* 2019;16: 235–46. https://doi.org/10.1038/s41575-018-0099-1

14. He, J., and Crews, F. T., "Increased MCP-1 and microglia in various regions of the human alcoholic brain," *Experimental Neurology* 2008; 210(2): 349–58. https://doi.org/10.1016/j.expneurol.2007.11.017

15. Piacentino, D., Vizioli, C., Barb, J. J., Grant-Beurmann, S., Bouhlal, S., Battista, J. T., Jennings, O., Lee, M. R., Schwandt, M. L., Walter, P., Henderson, W. A., Chen, K., Turner, S., Yang, S., Fraser, C. M., Farinelli, L. A., Farokhnia, M., and Leggio, L., "Gut microbial diversity and functional characterization in people with alcohol use disorder: A case-control study," *PLoS ONE* 2024;19(6): e0302195. https://doi.org/10.1371/journal.pone.0302195

16. Getachew, B., Hauser, S. R., Bennani, S., El Kouhen, N., Sari, Y., and Tizabi, Y., "Adolescent alcohol drinking interaction with the gut microbiome: Implications for adult alcohol use disorder," *Advances in Drug and Alcohol Research* 2024;4: 11881. https://doi.org/10.3389/adar.2024.11881

17. Bajaj, J. S., Gavis, E. A., Fagan, A., Wade, J. B., Thacker, L. R., Fuchs, M., Patel, S., Davis, B., Meador, J., Puri, P., Sikaroodi, M., and Gillevet, P. M., "A randomized clinical trial of fecal microbiota transplant for alcohol use disorder," *Hepatology* 2021;73(5): 1688–700. https://doi.org/10.1002/hep.31496

18. Antinozzi, M., Giffi, M., Sini, N., Gallè, F., Valeriani, F., De Vito, C., Liguori, G., Romano Spica, V., and Cattaruzza, M. S., "Cigarette smoking and human gut microbiota in healthy adults: A systematic review," *Biomedicines* 2022;10(2): 510. https://doi.org/10.3390/biomedicines10020510

19. Gui, X., Yang, Z., and Li, M. D., "Effect of cigarette smoke on gut microbiota: State of knowledge," *Frontiers in Physiology* 2021;12: 673341. https://doi.org/10.3389/fphys.2021.673341

20. Savin, Z., Kivity, S., Yonath, H., and Yehuda, S., "Smoking and the intestinal microbiome," *Archives in Microbiology* 2018;200(5): 677–84. https://doi.org/10.1007/s00203-018-1506-2

21. Talukder, M.A.H., Johnson, W. M., Varadharaj, S., Lian, J., Kearns, P. N., El-Mahdy, M. A., Liu, X., and Zweier, J. L., "Chronic cigarette smoking causes hypertension, increased oxidative stress, impaired NO bioavailability, endothelial dysfunction, and cardiac remodeling in mice," *American Journal of Physiology-Heart and Circulatory Physiology* 2011;300(1): H388–96. https://doi.org/10.1152/ajpheart.00868.2010

22. Arnson, Y., Shoenfeld, Y., and Amital, H., "Effects of tobacco smoke on immunity, inflammation and autoimmunity," *Journal of Autoimmunity* 2010;34(3): J258–65. https://doi.org/10.1016/j.jaut.2009.12.003

23. Tomoda, K., Kubo, K., Asahara, T., Andoh, A., Nomoto, K., Nishii, Y., Yamamoto, Y., Yoshikawa, M., and Kimura, H., "Cigarette smoke decreases organic acids levels and population of *bifidobacterium* in the caecum of rats," *Journal of Toxicological Sciences* 2011;36(3): 261–66. https://doi.org/10.2131/jts.36.261

24. Yang, H. T., Xiu, W.-J., Liu, J.-K., Yang, Y., Zhang, Y.-J., Zheng, Y.-Y., Wu, T.-T., Hou, X.-G., Wu, C.-X., Ma, Y.-T., and Xie, X., "Characteristics of the intestinal microorganisms in middle-aged and elderly patients: Effects of smoking," *ACS Omega* 2022;7(2): 1628–38. https://doi.org/10.1021/acsomega.1c02120

25. Yoon, H., Lee, D. H., Lee, J. H., Kwon, J. E., Shin, C. M., Yang, S.-J., Park, S.-H., Lee, J. H., Kang, S. W., Lee, J. S., and Kim, B.-Y., "Characteristics of the gut microbiome of healthy young male soldiers in South Korea: The effects of smoking," *Gut and Liver* 2021;15(2): 243–52. https://doi.org/10.5009/gnl19354

26. Jung, Y., Tagele, S. B., Son, H., Ibal, J. C., Kerfahi, D., Yun, H., Lee, B., Park, C. Y., Kim, E. S., Kim, S.-J., and Shin, J.-H., "Modulation of gut microbiota in Korean navy trainees following a healthy lifestyle change," *Microorganisms* 2020;8(9): 1265. https://doi.org/10.3390/microorganisms8091265

27. Fan, J., Zhou, Y., Meng, R., Tang, J., Zhu, J., Aldrich, M. C., Cox, N. J., Zhu, Y., Li, Y., and Zhou, D., "Cross-talks between gut microbiota and tobacco smoking: A two-sample Mendelian randomization study," *BMC Medicine* 2023;21(163). https://doi.org/10.1186/s12916-023-02863-1

28. Bercik, P., Park, A. J., Sinclair, D., Khoshdel, A., Lu, J., Huang, X., Deng, Y., Blennerhassett, P. A., Fahnestock, M., Moine, D., Berger, B., Huizinga, J. D., Kunze, W., McLean, P. G., Bergonzelli, G. E., Collins, S. M., and Verdu, E. F., "The anxiolytic effect of *Bifidobacterium longum* NCC3001 involves vagal pathways for gut-brain communication," *Neurogastroenterology & Motility* 2011;23(12): 1132–39. https://doi.org/10.1111/j.1365-2982.2011.01796.x

29. Han, W., Tellez, L. A., Perkins, M. H., Perez, I. O., Qu, T., Ferreira, J., Ferreira, T. L., Quinn, D., Liu, Z.-W., Gao, X.-B., Kaelberer, M. M., Bohórquez, D. V., Shammah-Lagnado, S. J., de Lartigue, G., and de Araujo, I. E., "A neural circuit for gut-induced reward," *Cell* 2018;175(3): 665–78.e623. https://doi.org/10.1016/j.cell.2018.08.049

30. Lee, L.-H., Goh, B.-H., and Chan, K.-G., "Editorial: Actinobacteria: Prolific producers of bioactive metabolites," *Frontiers in Microbiology* 2020;11: 1612. https://doi.org/10.3389/fmicb.2020.01612

31. Savin, Z., Kivity, S., Yonath, H., and Yehuda, S., "Smoking and the intestinal microbiome," *Archives in Microbiology* 2018;200(5): 677–84. https://doi.org/10.1007/s00203-018-1506-2

32. Sublette, M. G., Cross, T. -L., Korcarz, C. E., Hansen, K. M., Murga-Garrido, S. M., Hazen, S. L., Wang, Z., Oguss, M. K., Rey, F. E., and Stein, J. H., "Effects of smoking and smoking cessation on the intestinal microbiota," *Journal of Clinical Medicine* 2020;9(9): 2963. https://doi.org/10.3390/jcm9092963

33. Fluhr, L., Mor, U., Kolodziejczyk, A. A., Dori-Bachash, M., Leshem, A., Itav, S., Cohen, Y., Suez, J., Zmora, N., Moresi, C., Molina, S., Ayalon, N., Valdés-Mas, R., Hornstein, S., Karbi, H., Kviatcovsky, D., Livne, A., Bukimer, A., Eliyahu-Miller, S., Metz, A., Brandis, A., Mehlman, T., Kuperman, Y., Tsoory, M., Stettner, N., Harmelin, A., Shapiro, H., and Elinav, E., "Gut microbiota modulates weight gain in mice after discontinued smoke exposure," *Nature* 2021;600: 713–19. https://doi.org/10.1038/s41586-021-04194-8

34. Ohue-Kitano, R., Banno, Y., Masujima, Y., and Kimura, I., "Gut microbial metabolites reveal diet-dependent metabolic changes induced by nicotine administration," *Scientific Reports* 2024;14: 1056. https://doi.org/10.1038/s41598-024-51528-3

35. Sohail, R., Mathew, M., Patel, K. K., Reddy, S. A., Haider, Z., Naria, M., Habib, A., Abdin, Z. U., Razzaq Chaudhry, W., and Akbar, A., "Effects of non-steroidal anti-inflammatory drugs (NSAIDs) and gastroprotective NSAIDs on the gastrointestinal tract: A narrative review," *Cureus* 2023;15(4): e37080. https://doi.org/10.7759/cureus.37080

36. Mayo, S. A., Song, Y. K., Cruz, M. R., Phan, T. M., Singh, K. V., Garsin, D. A., Murray, B. E., Dial, E. J., and Lichtenberger, L. M., "Indomethacin injury to the rat small intestine is dependent upon biliary secretion and is associated with overgrowth of enterococci," *Physiological Reports* 2016;4(6): e12725. https://doi.org/10.14814/phy2.12725

37. RSNA Press Release: "NSAIDs May Worsen Arthritis Inflammation,"

November 21, 2022. https://press.rsna.org/timssnet/media/pressreleases /14_pr_target.cfm?id=2379

38. Cruz-Lebron, A., Johnson, R., Mazahery, C., Troyer, Z., Joussef-Piña, S., Quiñones-Mateu, M. E., Strauch, C. M., Hazen, S. L., and Levine, A. D., "Chronic opioid use modulates human enteric microbiota and intestinal barrier integrity," *Gut Microbes* 2021;13(1): 1946368. https://doi.org/10 .1080/19490976.2021.1946368

39. Zádori, Z. S., Király, K., Al-Khrasani, M., and Gyires, K., "Interactions between NSAIDs, opioids and the gut microbiota—future perspectives in the management of inflammation and pain," *Pharmacology & Therapeutics* 2023;241: 108327. https://doi.org/10.1016/j .pharmthera.2022.108327

40. Lee, K., Vuong, H. E., Nusbaum, D. J., Hsiao, E. Y., Evans, C. J., and Taylor, A.M.W., "The gut microbiota mediates reward and sensory responses associated with regimen-selective morphine dependence," *Neuropsychopharmacology* 2018;43: 2606–14. https://doi.org/10.1038/s41386-018 -0211-9

41. Wang, F., and Roy, S., "Gut homeostasis, microbial dysbiosis, and opioids," *Toxicologic Pathology* 2017;45(1): 150–56. https://doi.org/10.1177 /0192623316679898

42. Wang, F., Meng, J., Zhang, L., Johnson, T., Chen, C., and Roy, S., "Morphine induces changes in the gut microbiome and metabolome in a morphine dependence model," *Scientific Reports* 2018;8(1): 3596. https://doi .org/10.1038/s41598-018-21915-8

43. Zhang, L., Meng, J., Ban, Y., Jalodia, R., Chupikova, I., Fernandez, I., Brito, N., Sharma, U., Abreu, M. T., Ramakrishnan, S., and Roy, S., "Morphine tolerance is attenuated in germfree mice and reversed by probiotics, implicating the role of gut microbiome," *Proceedings of the National Academy of Sciences USA* 2019;116(27): 13523–32. https://doi.org/10.1073 /pnas.1901182116

44. Rivat, C., and Ballantyne, J., "The dark side of opioids in pain management: Basic science explains clinical observation," *PAIN Reports* 2016 1(2): e570. https://doi.org/10.1097/PR9.0000000000000570

45. Lewin-Epstein, O., Jaques, Y., Feldman, M. W., Kaufer, D., and Hadany, L., "Evolutionary modeling suggests that addictions may be driven by competition-induced microbiome dysbiosis," *Communications Biology* 2023;6(782). https://doi.org/10.1038/s42003-023-05099-0

46. Molavi, N., Rasouli-Azad, M., Mirzaei, H., Matini, A. H., Banafshe, H. R., Valiollahzadeh, M., Hassanzadeh, M., Saghazade, A. R.,

Abbaszadeh-Mashkani, S., Mamsharifi, P., and Ghaderi, A., "The effects of probiotic supplementation on opioid-related disorder in patients under methadone maintenance treatment programs," *International Journal of Clinical Practice* 2022;30: 1206914. https://doi .org/10.1155/2022/1206914

CHAPTER 7: MENTAL HEALTH AND THE GUT

1. Bettencourt Rodrigues, A. M., "De l'influence des phénomènes d'auto-intoxication et de la dilatation de l'estomac dans les formes dépressives et mélancoliques" (extrait des comptes rendus du congrès de médecine mentale, tenu à Paris du 5 au 10 août 1889): 5, 7.
2. Ernst, E., "Colonic irrigation and the theory of autointoxication: A triumph of ignorance over science," *Journal of Clinical Gastroenterology* 1997;24(4):196–98. https://doi.org/10.1097/00004836-199706000-00002
3. Mathias, M., "Autointoxication and historical precursors of the microbiome-gut-brain axis," *Microbial Ecology in Health and Disease* 2018;29(2): 1548249. https://doi.org/10.1080/16512235.2018.1548249.
4. Chevalier-Lavaure, F-A., "Des Auto-intoxications dans les maladies mentales, Contribution à l'Etude de la pathogénie de la folie [thèse Bordeaux]" (soutenue 30 juillet 1890): 91, 25, 86, 87, 88.
5. Pearson-Leary, J., Zhao, C., Bittinger, K., Eacret, D., Luz, S., Vigderman, A. S., Dayanim, G., and Bhatnagar, S., "The gut microbiome regulates the increases in depressive-type behaviors and in inflammatory processes in the ventral hippocampus of stress vulnerable rats," *Molecular Psychiatry* 2020;25(5): 1068–79. https://doi.org/10.1038/s41380-019-0380-x
6. Pape, K., Tamouza, R., Leboyer, M., and Zipp, F., "Immunoneuropsychiatry—novel perspectives on brain disorders," *Nature Reviews Neurology* 2019;15(6): 317–28. https://doi.org/10.1038/s41582-019-0174-4
7. Medina-Rodriguez, E. M., Madorma, D., O'Connor, G., Mason, B. L., Han, D., Deo, S. K., Oppenheimer, M., Nemeroff, C. B., Trivedi, M. H., Daunert, S., and Beurel, E., "Identification of a signaling mechanism by which the microbiome regulates Th17 cell-mediated depressive-like behaviors in mice," *American Journal of Psychiatry* 2020;177(10): 974–90. https://doi.org/10.1176/appi.ajp.2020.19090960
8. Ivanov, I. I., Tuganbaev, T., Skelly, A. N., and Honda, K., "T cell responses to the microbiota," *Annual Review of Immunology* 2022;40: 559–87. https://doi.org/10.1146/annurev-immunol-101320-011829
9. Fields, C. T., Chassaing, B., Castillo-Ruiz, A., Osan, R., Gewirtz, A. T.,

and de Vries, G. J., "Effects of gut-derived endotoxin on anxiety-like and repetitive behaviors in male and female mice," *Biology of Sex Differences* 2018;9(7). https://doi.org/10.1186/s13293-018-0166-x

10. Kageyama, Y., Kasahara, T., Kato, M., Sakai, S., Deguchi, Y., Tani, M., Kuroda, K., Hattori, K., Yoshida, S., Goto, Y., Kinoshita, T., Inoue, K., and Kato, T., "The relationship between circulating mitochondrial DNA and inflammatory cytokines in patients with major depression," *Journal of Affective Disorders* 2018;233: 15–20. https://doi.org/10.1016/j.jad.2017.06.001

11. Lindqvist, D., Fernström, J., Grudet, C., Ljunggren, L., Träskman-Bendz, L., Ohlsson, L., and Westrin, Å, "Increased plasma levels of circulating cell-free mitochondrial DNA in suicide attempters: Associations with HPA-axis hyperactivity," *Translational Psychiatry* 2016; 6(12): e971. https://doi.org/10.1038/tp.2016.236

12. Wicherska-Pawłowska, K., Wróbel, T., and Rybka, J., "Toll-like receptors (TLRs), NOD-like receptors (NLRs), and RIG-I-like receptors (RLRs) in innate immunity. TLRs, NLRs, and RLRs ligands as immunotherapeutic agents for hematopoietic diseases," *International Journal of Molecular Sciences* 2021;22(24): 13397. https://doi.org/10.3390/ijms222413397

13. Gambardella, S., Limanaqi, F., Ferese, R., Biagioni, F., Campopiano, R., Centonze, D., and Fornai, F., "ccf-mtDNA as a potential link between the brain and immune system in neuro-immunological disorders," *Frontiers in Immunology* 2019;10: 1064. https://doi.org/10.3389/fimmu.2019.01064

14. Goodwin, R. D., Dierker, L. C., Wu, M., Galea, S., Hoven, C. W., and Weinberger, A. H., "Trends in U.S. depression prevalence from 2015 to 2020: The widening treatment gap," *American Journal of Preventive Medicine* 2022;63(5): 726–33. https://doi.org/10.1016/j.amepre.2022.05.014

15. Radjabzadeh, D., Bosch, J. A., Uitterlinden, A. G., Zwinderman, A. H., Ikram, M. A., van Meurs, J.B.J., Luik, A. I., Nieuwdorp, M., Lok, A., van Duijn, C. M., Kraaij, R., and Amin, N., "Gut microbiome-wide association study of depressive symptoms," *Nature Communications* 2022;13(1): 7128. https://doi.org/10.1038/s41467-022-34502-3

16. Liu, L., Wang, H., Zhang, H., Chen, X., Zhang, Y., Wu, J., Zhao, L., Wang, D., Pu, J., Ji, P., and Xie, P., "Toward a deeper understanding of gut microbiome in depression: The promise of clinical applicability," *Advanced Science (Weinh)* 2022;9(35): e2203707. https://doi.org/10.1002/advs.202203707

17. Liu, L., Wang, H., Zhang, H., Chen, X., Zhang, Y., Wu, J., Zhao, L., Wang, D., Pu, J., Ji, P., and Xie, P., "Toward a deeper understanding of gut microbiome in depression: The promise of clinical applicability," *Advanced Science (Weinh)* 2022;9(35): e2203707. https://doi.org/10.1002/advs.202203707

18. Borkent, J., Ioannou, M., Laman, J. D., Haarman, B.C.M., and Sommer, I.E.C., "Role of the gut microbiome in three major psychiatric disorders," *Psychological Medicine* 2022;52(7): 1222–42. https://doi.org/10.1017/S0033291722000897

19. Wilkowska, A., Szałach. Ł.P., and Cubała, W. J., "Gut microbiota in depression: A focus on ketamine," *Frontiers in Behavioral Neuroscience* 2021;15: 693362. https://doi.org/10.3389/fnbeh.2021.693362

20. Radjabzadeh, D., Bosch, J. A., Uitterlinden, A. G., Zwinderman, A. H., Ikram, M. A., van Meurs, J.B.J., Luik, A. I., Nieuwdorp, M., Lok, A., van Duijn, C. M., Kraaij, R., and Amin N., "Gut microbiome-wide association study of depressive symptoms," *Nature Communications* 2022;13(1): 7128. https://doi.org/10.1038/s41467-022-34502-3

21. Safadi, J. M., Quinton, A.M.G., Lennox, B. R., Burnet, P.W.J., and Minichino, A., "Gut dysbiosis in severe mental illness and chronic fatigue: A novel trans-diagnostic construct? A systematic review and meta-analysis," *Molecular Psychiatry* 2022;27: 141–53. https://doi.org/10.1038/s41380-021-01032-1

22. D'Mello, C., and Swain, M. G., "Liver-brain interactions in inflammatory liver diseases: Implications for fatigue and mood disorders," *Brain, Behavior, and Immunity* 2014;35: 9–20. https://doi.org/10.1016/j.bbi.2013.10.009

23. Erabi, H., Okada, G., Shibasaki, C., Setoyama, D., Kang, D., Takamura, M., Yoshino, A., Fuchikami, M., Kurata, A., Kato, T. A., Yamawaki, S., and Okamoto, Y., "Kynurenic acid is a potential overlapped biomarker between diagnosis and treatment response for depression from metabolome analysis," *Scientific Reports* 2020;10: 16822. https://doi.org/10.1038/s41598-020-73918-z

24. Gavrilova, S. I., Kolykhalov, I. V., Fedorova, Ia.B., Selezneva, N. D., Kalyn, Ia.B., Roshchina, I. F., Odinak, M. M., Emelin, Iu.A., Kashin, A. V., Gustov, A. V., Antipenko, E. A., Korshunova, Iu.A., Davydova, T. A., and Messler, G., "Possibilities of preventive treatment of Alzheimer's disease: Results of the 3-year open prospective comparative study on efficacy and safety of the course therapy with cerebrolysin and cavinton in elderly patients with the syndrome of mild cognitive impairment," *Zhurnal Nevrologii i Psikhiatrii Imeni S. S. Korsakova* 2010;110(1):62–69. https://pubmed.ncbi.nlm.nih.gov/20436440/

25. Zimmermann, M., Zimmermann-Kogadeeva, M., Wegmann, R., and Goodman, A. L., "Mapping human microbiome drug metabolism by gut bacteria and their genes," *Nature* 2019;570: 462–67. https://doi.org/10.1038/s41586-019-1291-3

26. Shen, Y., Yang, X., Li, G., Gao, J., and Liang, Y., "The change of gut microbiota in MDD patients under SSRIs treatment," *Scientific Reports* 2021;11: 14918. https://doi.org/10.1038/s41598-021-94481-1

27. Chinna Meyyappan, A., Forth, E., Wallace, C.J.K., and Milev, R., "Effect of fecal microbiota transplant on symptoms of psychiatric disorders: A systematic review," *BMC Psychiatry* 2020;20(299). https://doi.org/10.1186/s12888-020-02654-5

28. Angeletti, S., Dicuonzo, G., Lo Presti, A., Cella, E., Crea, F., Avola, A., Vitali, M. A., Fagioni, M., and De Florio, L., "MALDI-TOF mass spectrometry and blakpc gene phylogenetic analysis of an outbreak of carbapenem-resistant K. pneumoniae strains," *New Microbiologica* 2015;38(4): 541–50. https://pubmed.ncbi.nlm.nih.gov/26485012/

29. Burokas, A., Arboleya, S., Moloney, R. D., Peterson, V. L., Murphy, K., Clarke, G., Stanton, C., Dinan, T. G., and Cryan, J. F., "Targeting the microbiota-gut-brain axis: Prebiotics have anxiolytic and antidepressant-like effects and reverse the impact of chronic stress in mice," *Biological Psychiatry* 2017;82(7): 472–87. https://doi.org/10.1016/j.biopsych.2016.12.031

30. Bistas, K. G., and Tabet, J. P., "The benefits of prebiotics and probiotics on mental health," *Cureus* 2023;15(8): e43217. https://doi.org/10.7759/cureus.43217

31. Bao, P., Gong, Y., Wang, Y., Xu, M., Qian, Z., Ni, X., and Lu, J., "Hydrogen sulfide prevents LPS-induced depression-like behavior through the suppression of NLRP3 inflammasome and pyroptosis and the improvement of mitochondrial function in the hippocampus of mice," *Biology* 2023;12(8): 1092. https://doi.org/10.3390/biology12081092

32. Woller, S. A., Eddinger, K. A., Corr, M., and Yaksh, T. L., "An overview of pathways encoding nociception," *Clinical and Experimental Rheumatology* 2017;35 Suppl. 107(5):40–46. https://pubmed.ncbi.nlm.nih.gov/28967373/

33. Zhang, W., Lyu, M., Dessman, N. J., Xie, Z., Arifuzzaman, M., Yano, H., Parkhurst, C. N., Chu, C., Zhou, L., Putzel, G. G., Li, T. T., Jin, W.-B., Zhou, J., JRI Live Cell Bank, Hu, H., Tsou, A. M., Guo, C.-J., and Artis, D., "Gut-innervating nociceptors regulate the intestinal microbiota to promote tissue protection," *Cell* 2022;185(22): 4170–89.e20. https://doi.org/10.1016/j.cell.2022.09.008

34. Szabo, C., and Papapetropoulos, A., "International union of basic and clinical pharmacology. CII: Pharmacological modulation of H_2S levels: H_2S donors and H_2S biosynthesis inhibitors," *Pharmacological Reviews* 2017;69(4): 497–564. https://doi.org/10.1124/pr.117.014050

35. Hou, X.-Y., Hu, Z.-L., Zhang, D.-Z., Lu, W., Zhou, J., Wu, P.-F., Guan, X.-L., Han, Q.-Q., Deng, S.-L., Zhang, H., Chen, J.-G., and Wang, F., "Rapid antidepressant effect of hydrogen sulfide: Evidence for activation of mTORC1-TrkB-AMPA receptor pathways," *Antioxidants & Redox Signaling* 2017;27(8): 472–88. https://doi.org/10.1089/ars.2016.6737

36. Sala-Cirtog, M., and Sirbu, I.-O., "Analysis of microRNA-transcription factors co-regulatory network linking depression and vitamin D deficiency," *International Journal of Molecular Sciences* 2024;25(2): 1114. https://doi.org/10.3390/ijms25021114

37. Thomas, R. L., Jiang, L., Adams, J. S., Xu, Z. Z., Shen, J., Janssen, S., Ackermann, G., Vanderschueren, D., Pauwels, S., Knight, R., Orwoll, E. S., and Kado, D., "Vitamin D metabolites and the gut microbiome in older men," *Nature Communications* 2020;11: 5997. https://doi.org/10.1038/s41467-020-19793-8

38. Singh, P., Rawat, A., Alwakeel, M., Sharif, E., and Al Khodor, S., "The potential role of vitamin D supplementation as a gut microbiota modifier in healthy individuals," *Scientific Reports* 2020;10(1): 21641. https://doi.org/10.1038/s41598-020-77806-4

39. Hahn, J., Cook, N. R., Alexander, E. K., Friedman, S., Walter, J., Bubes, V., Kotler, G., Lee, I. M., Manson, J. E., and Costenbader, K. H., "Vitamin D and marine omega 3 fatty acid supplementation and incident autoimmune disease: VITAL randomized controlled trial," *BMJ* 2022;376: e066452. https://doi.org/10.1136/bmj-2021-066452

40. Ghahremani, M., Smith, E. E., Chen, H.-Y., Creese, B., Goodarzi, Z., and Ismail, Z., "Vitamin D supplementation and incident dementia: Effects of sex, *APOE*, and baseline cognitive status," *Alzheimer's & Dementia* 2023;15(1): e12404. https://doi.org/10.1002/dad2.12404

41. Szałach, Ł. P., Lisowska, K. A., Słupski, J., Włodarczyk, A., Górska, N., Szarmach J., Jakuszkowiak-Wojten, K., Gałuszko-Węgielnik, M., Wiglusz, M. S., Wiklowska, A., and Cubała, W. J., "The immunomodulatory effect of ketamine in depression," *Psychiatria Danubina* 2019;31 (Suppl. 3): 252–57. https://pubmed.ncbi.nlm.nih.gov/31488736/

42. Getachew, B., Aubee, J. I., Schottenfeld, R. S., Csoka, A. B., Thompson, K. M., and Tizabi, Y., "Ketamine interactions with gut-microbiota in rats: Relevance to its antidepressant and anti-inflammatory properties," *BMC Microbiology* 2018;18: 222. https://doi.org/10.1186/s12866-018-1373-7

43. Verdonk, F., Petit, A.-C., Abdel-Ahad, P., Vinckier, F., Jouvion, G., de Maricourt, P., De Medeiros, G. F., Danckaert, A., Van Steenwinckel, J., Blatzer, M., Maignan, A., Langeron, O., Sharshar, T., Callebert, J., Launay, J.-M., Chrétien, F., and Gaillard, R., "Microglial production of quinolinic acid as a target and a biomarker of the antidepressant effect of ketamine," *Brain, Behavior, and Immunity* 2019;81: 361–73. https://doi .org/10.1016/j.bbi.2019.06.033

44. Getachew, B., Aubee, J. I., Schottenfeld, R. S., Csoka, A. B., Thompson, K. M., and Tizabi, Y., "Ketamine interactions with gut-microbiota in rats: Relevance to its antidepressant and anti-inflammatory properties," *BMC Microbiology* 2018(18): 222. https://doi.org/10.1186/s12866-018-1373-7

45. Dinis-Oliveira, R. J., "Metabolism of psilocybin and psilocin: Clinical and forensic toxicological relevance," *Drug Metabolism Reviews* 2017;49(1): 84–91. https://doi.org/10.1080/03602532.2016.1278228

46. Lenz, C., Sherwood, A., Kargbo, R., and Hoffmeister, D., "Taking different roads: L-tryptophan as the origin of *Psilocybe* natural products," *ChemPlusChem* 2021;86(1): 28–35. https://doi.org/10.1002 /cplu.202000581

47. Kelly, J. R., Clarke, G., Harkin, A., Corr, S. C., Galvin, S., Pradeep, V., Cryan, J. F., O'Keane, V., and Dinan, T. G., "Seeking the psilocybiome: Psychedelics meet the microbiota-gut-brain axis," *International Journal of Clinical and Health Psychology* 2023;23(2): 100349. https:// doi.org/10.1016/j.ijchp.2022.100349

48. López-Giménez, J. F., and González-Maeso, J., "Hallucinogens and serotonin 5-HT$_{2A}$ receptor-mediated signaling pathways," *Current Topics in Behavioral Neurosciences* 2018; 36: 45–73. https://doi .org/10.1007/7854_2017_478

49. National Institutes of Health, NIH Research Matters, "How psychedelic drugs may help with depression," March 14, 2023. https://www.nih.gov /news-events/nih-research-matters/how-psychedelic-drugs-may-help -depression

50. Mezquita, G. A., Correia, B. L., Bastos, I. M., Bentes, A. S., and de Carvalho, J. G., "The missing link: Does the gut microbiota influence the efficacy of psychedelics?" *Journal of Psychopharmacology* 2021;35: 495–507.

51. Szabo, A., Kovacs, A., Riba, J., Djurovic, S., Rajnavolgyi, E., and Frecska, E., "The endogenous hallucinogen and trace amine N,N-dimethyltryptamine (DMT) displays potent protective effects against hypoxia via sigma-1 receptor activation in human primary iPSC-derived cortical neurons and microglia-like immune cells," *Frontiers in Neuroscience* 2016;10: 423. https://doi.org/10.3389/fnins.2016.00423

52. Nichols, D. E., "Psychedelics," *Pharmacological Reviews* 2016;68(2): 264–355. https://doi.org/10.1124/pr.115.011478

53. Saito, Y., Hashimoto, K., Cyranoski, D., Yokoyama, A., Suzuki, K., Kawase, Y., Takeda, T., and Hongo, S., "The gut microbiota and psychiatric disorders: Implications for clinical practice," *Journal of Clinical Psychiatry* 2019;80: 1–11.

54. Nichols, D. E., "Psychedelics," *Pharmacological Reviews* 2016;68(2): 264–355. https://doi.org/10.1124/pr.115.011478

55. Wu, G. D., Chen, J., Hoffmann, C., Bittinger, K., Chen, Y.-Y., Keilbaugh, S. A., Bewtra, M., Knights, D., Walters, W. A., Knight, R., Sinha, R., Gilroy, E., Gupta, K., Baldassano, R., Nessel, L., Li, H., Bushman, F. D., and Lewis, J. D., "Linking long-term dietary patterns with gut microbial enterotypes," *Science* 2011; 334(6052): 105–08. https://doi.org/10.1126/science.1208344

56. DeNapoli, J. S., Dodman, N. H., Shuster, L., Rand, W. M., and Gross, K. L., "Effect of dietary protein content and tryptophan supplementation on dominance aggression, territorial aggression, and hyperactivity in dogs," *Journal of the American Veterinary Medical Association* 2000;217(4): 504–08. https://doi.org/10.2460/javma.2000.217.504

57. Malekahmadi, M., Khayyatzadeh, S. S., Heshmati, J., Alshahrani, S. H., Oraee, N., Ferns, G. A., Firouzi, S., Pahlavani, N., and Ghayour-Mobarhan, M., "The relationship between dietary patterns and aggressive behavior in adolescent girls: A cross-sectional study," *Brain and Behavior* 2022;12(12): e2782. https://doi.org/10.1002/brb3.2782

58. Shin, J.-H., Kim, C.-S., Cha, J., Sojeong, K., Lee, S., Chae, S., Chun, W. Y., and Shin, D.-M., "Consumption of 85% cocoa dark chocolate improves mood in association with gut microbial changes in healthy adults: A randomized controlled trial," *Journal of Nutritional Biochemistry* 2022;99: 108854. https://doi.org/10.1016/j.jnutbio.2021.108854

59. Rezzi, S., Ramadan, Z., Martin, F.-P.J., Fay, L. B., van Bladeren, P., Lindon, J. C., Nicholson, J. K., and Kochhar, S., "Human metabolic phenotypes link directly to specific dietary preferences in healthy individuals," *Journal of Proteome Research* 2007;6(11): 4469–77. https://doi.org/10.1021/pr070431h

60. Sarr, M. G., Billington, C. J., Brancatisano, R., Brancatisano, A., Toouli, J., Kow, L., Nguyen, N. T., Blackstone, R., Maher, J. W., Shikora, S., Reeds, D. N., Eagon, J. C., Wolfe, B. M., O'Rourke, R. W., Fujioka, K., Takata, M., Swain, J. M., Morton, J. M., Ikramuddin, S., Schweitzer, M., Chand, B., Rosenthal, P., and EMPOWER Study Group, "The EMPOWER study: Randomized, prospective, double-blind, multicenter trial of vagal

blockade to induce weight loss in morbid obesity," *Obesity Surgery* 2012;22(11): 1771–82. https://doi.org/10.1007/s11695-012-0751-8

61. Camilleri, M., Toouli, J., Herrera, M. F., Kulseng, B., Kow, L., Pantoja, J. P., Marvik, R., Johnsen, G., Billington, C. J., Moody, F. G., Knudson, M. B., Tweden, K. S., Vollmer, M., Wilson, R. R., and Anvari, M., "Intra-abdominal vagal blocking (VBLOC therapy): Clinical results with a new implantable medical device," *Surgery* 2008;143(6): 723–31. https://doi .org/10.1016/j.surg.2008.03.015

62. Kollai, M., Bonyhay, I., Jokkel, G., and Szonyi, L., "Cardiac vagal hyperactivity in adolescent anorexia nervosa," *European Heart Journal*, 1994;15(8): 1113–18. https://doi.org/10.1093/oxfordjournals.eurheartj.a060636

63. Duca, F. A., Swartz, T. D., Sakar, Y., and Covasa, M., "Increased oral detection, but decreased intestinal signaling for fats in mice lacking gut microbiota," *PLoS ONE* 2012;7(6): e39748. https://doi.org/10.1371 /journal.pone.0039748

64. Swartz, T. D., Duca, F. A., de Wouters, T., Sakar, Y., and Covasa, M., "Up-regulation of intestinal type 1 taste receptor 3 and sodium glucose luminal transporter-1 expression and increased sucrose intake in mice lacking gut microbiota," *British Journal of Nutrition* 2012;107(5): 621–30. https://doi .org/10.1017/S0007114511003412

65. Duca, F. A., Swartz, T. D., Sakar, Y., and Covasa, M., "Increased oral detection, but decreased intestinal signaling for fats in mice lacking gut microbiota," *PLoS ONE* 2012;7(6): e39748. https://doi.org/10.1371 /journal.pone.0039748

66. Fetissov, S. O., Hamze Sinno, M., Coëffier, M., Bole-Feysot, C., Ducrotté, P., Hökfelt, T., and Déchelotte, P., "Autoantibodies against appetite-regulating peptide hormones and neuropeptides: Putative modulation by gut microflora," *Nutrition* 2008;24(4): 348–59. https://doi.org/10.1016/j .nut.2007.12.006

67. Wisse, B. E., Ogimoto, K., Tang, J., Harris, M. K., Jr., Raines, E. W., and Schwartz, M. W., "Evidence that lipopolysaccharide-induced anorexia depends upon central, rather than peripheral, inflammatory signals," *Endocrinology* 2007;148(11): 5230–37. https://doi.org/10.1210 /en.2007-0394

68. Broft, A., Shingleton, R., Kaufman, J., Liu, F., Kumar, D., Slifstein, M., Abi-Dargham, A., Schebendach, J., Van Heertum, R., Attia, E., Martinez, D., and Walsh, B. T., "Striatal dopamine in bulimia nervosa: A PET imaging study," *International Journal of Eating Disorders* 2012;45(5): 648–56. https://doi.org/10.1002/eat.20984

69. Zink, C. F., and Weinberger, D. R., "Cracking the moody brain: The rewards of self starvation," *Nature Medicine* 2010;16(12): 1382–83. https://doi.org/10.1038/nm1210-1382

70. Herman, A., and Bajaka, A., "The role of the intestinal microbiota in eating disorders—bulimia nervosa and binge eating disorder," *Psychiatry Research* 2021;300: 113923. https://doi.org/10.1016/j.psychres.2021.113923

71. Kaye, W. H., Fudge, J. L., and Paulus, M., "New insights into symptoms and neurocircuit function of anorexia nervosa," *Nature Reviews Neuroscience* 2009;10: 573–84. https://doi.org/10.1038/nrn2682

CHAPTER 8: NEURO-INFLAMM-AGING AND THE GUT

1. Franceschi, C., Bonafè, M., Valensin, S., Olivieri, F., De Luca, M., Ottaviani, E., and De Benedictis, G., "Inflamm-aging: An evolutionary perspective on immunosenescence," *Annals of the New York Academy of Sciences* 2000;908(1): 244–54. https://doi.org/10.1111/j.1749-6632.2000.tb06651.x

2. Minhoo, K., and Benayoun, B. A., "The microbiome: An emerging key player in aging and longevity," *Translational Medicine of Aging* 2020; 4: 103–16, https://doi.org/10.1016/j.tma.2020.07.004.

3. Kousparou, C., Fyrilla, M., Stephanou, A., and Patrikios, I., "DHA/EPA (omega-3) and LA/GLA (omega-6) as bioactive molecules in neurodegenerative diseases," *International Journal of Molecular Sciences* 2023;24(13): 10717. https://doi.org/10.3390/ijms241310717

4. Dai, C.-L., Liu, F., Iqbal, K., and Gong, C.-X., "Gut microbiota and immunotherapy for Alzheimer's disease," *International Journal of Molecular Sciences* 2022;23(23): 15230. https://doi.org/10.3390/ijms232315230

5. Yang, Y., Jiang, G., Zhang, P., and Fan, J., "Programmed cell death and its role in inflammation," *Military Medical Research* 2015;2(12). https://doi.org/10.1186/s40779-015-0039-0

6. Selkoe, D. J., "Folding proteins in fatal ways," *Nature* 2003;426: 900–04. https://doi.org/10.1038/nature02264

7. Sitia, R., and Braakman, I., "Quality control in the endoplasmic reticulum protein factory," *Nature* 2003;426(6968): 891–94. https://doi.org/10.1038/nature02262

8. Taylor, J. P., Hardy, J., and Fischbeck, K. H., "Toxic proteins in neurodegenerative disease," *Science* 2002;296(5575): 1991–95. https://doi.org/10.1126/science.1067122

9. Wang, C., Lau, C. Y., Ma, F., and Zheng, C., "Genome-wide screen identifies curli amyloid fibril as a bacterial component promoting host

neurodegeneration," *Proceedings of the National Academy of Sciences USA* 2021;118(34): e2106504118. https://doi.org/10.1073/pnas.2106504118

10. University of Birmingham, "Link between inflammation and mental sluggishness shown in new study," *ScienceDaily*, November 15, 2019 (retrieved May 7, 2024). www.sciencedaily.com/releases/2019/11/191115190337.htm

11. Biagi, E., Nylund, L., Candela, M., Ostan, R. O., Bucci, L., Pini, E., Nikkïla, J., Monti, D., Satokari, R., Franceschi, C., Brigidi, P., and De Vos, W., "Through ageing, and beyond: Gut microbiota and inflammatory status in seniors and centenarians," *PLoS ONE* 2010;5(5): e10667. https://doi.org/10.1371/journal.pone.0010667

12. Hou, Q., Ye, L., Liu, H., Huang, L., Yang, Q., Turner, J. R., and Yu, Q., "*Lactobacillus* accelerates ISCs regeneration to protect the integrity of intestinal mucosa through activation of STAT3 signaling pathway induced by LPLs secretion of IL-22," *Cell Death & Differentiation* 2018;25(9): 1657–70. https://doi.org/10.1038/s41418-018-0070-2

13. Mokkala, K., Laitinen, K., and Röytiö, H., "*Bifidobacterium lactis* 420 and fish oil enhance intestinal epithelial integrity in Caco-2 cells," *Nutrition Research* 2016;36(3): 246–52. https://doi.org/10.1016/j.nutres.2015.11.014

14. Ling, X., Linglong, P., Weixia, D., and Hong, W., "Protective effects of Bifidobacterium on intestinal barrier function in LPS-induced enterocyte barrier injury of Caco-2 monolayers and in a rat NEC model," *PLoS ONE* 2016;11(8): e0161635. https://doi.org/10.1371/journal.pone.0161635

15. Geerlings, S. Y., Kostopoulos, I., de Vos, W., and Belzer, C., "*Akkermansia muciniphila* in the human gastrointestinal tract: When, where, and how?" *Microorganisms* 2018;6(3): 75. https://doi.org/10.3390/microorganisms6030075

16. Norwitz, N. G., Saif, N., Ariza, I. E., and Isaacson, R. S., "Precision nutrition for Alzheimer's prevention in *ApoE4* carriers," *Nutrients* 2021;13(4): 1362. https://doi.org/10.3390/nu13041362

17. Miklossy, J., "Alzheimer's disease—a neurospirochetosis. Analysis of the evidence following Koch's and Hill's criteria," *Journal of Neuroinflammation* 2011;8: 90. https://doi.org/10.1186/1742-2094-8-90

18. Kumar, D.K.V., Choi, S. H., Washicosky, K. J., Eimer, W. A., Tucker, S., Ghofrani, J., Lefkowitz, A., McColl, G., Golstein, L. E., Tanzi, R., and Moir, R., "Amyloid-β peptide protects against microbial infection in mouse and worm models of Alzheimer's disease," *Science Translational Medicine* 2016;8(340): 340ra72. https://doi.org/10.1126/scitranslmed.aaf1059

19. Liu, S., Gao, J., Zhu, M., Liu, K., and Zhang, H.-L., "Gut microbiota and dysbiosis in Alzheimer's disease: Implications for pathogenesis and treatment," *Molecular Neurobiology* 2020;57: 5026–43. https://doi .org/10.1007/s12035-020-02073-3

20. Vojtechova, I., Machacek, T., Kristofikova, Z., Stuchlik, A., and Petrasek, T., "Infectious origin of Alzheimer's disease: Amyloid beta as a component of brain antimicrobial immunity," *PLoS Pathogens* 2022;18(11): e1010929. https://doi.org/10.1371/journal.ppat.1010929

21. Friedland, R. P., and Chapman, M. R., "The role of microbial amyloid in neurodegeneration," *PLoS Pathogens* 2017;13(12): e1006654. https://doi .org/10.1371/journal.ppat.1006654

22. Walker, A. C., Bhargava, R., Vaziriyan-Sani, A. S., Pourciau, C., Donahue, E. T., Dove, A. S., Gebhardt, M. J., Ellward, G. L., Romeo, T., and Czyż, D. M., "Colonization of the *Caenorhabditis elegans* gut with human enteric bacterial pathogens leads to proteostasis disruption that is rescued by butyrate," *PLoS Pathogens* 2021;17(5): e1009510. https://doi.org/10.1371 /journal.ppat.1009510

23. Cherny, I., Rockah, L., Levy-Nissenbaum, O., Gophna, U., Ron, E. Z., and Gazit, E., "The formation of *Escherichia coli* curli amyloid fibrils is mediated by prion-like peptide repeats," *Journal of Molecular Biology* 2005;352(2): 245–52. https://doi.org/10.1016/j.jmb.2005.07.028

24. Friedland, R. P., and Chapman, M. R., "The role of microbial amyloid in neurodegeneration," *PLoS Pathog* 2017;13(12): e1006654. https://doi .org/10.1371/journal.ppat.1006654

25. Zhao, Y., Dua, P., and Lukiw, W. J., "Microbial sources of amyloid and relevance to amyloidogenesis and Alzheimer's disease (AD)," *Journal of Alzheimers Disease & Parkinsonism* 2015;5(1): 177. https://doi .org/10.4172/2161-0460

26. Friedland, R. P., "Mechanisms of molecular mimicry involving the microbiota in neurodegeneration," *Journal of Alzheimer's Disease* 2015; 45(2): 349–62. https://doi.org/10.3233/JAD-142841

27. Friedland, R. P., "Mechanisms of molecular mimicry involving the microbiota in neurodegeneration," *Journal of Alzheimer's Disease* 2015;45(2): 349–62. https://doi.org/10.3233/JAD-142841

28. Itzhaki, R. F., Woan-Ru, L., Shang, D., Wilcock, G. K., Faragher, B., and Jamieson, G. A., "Herpes simplex virus type 1 in brain and risk of Alzheimer's disease," *Lancet* 1997;349(9047): 241–44. https://doi .org/10.1016/S0140-6736(96)10149-5

29. Gérard, H. C., Dreses-Werringloer, U., Wildt, K. S., Deka, S., Oszust, C., Balin, B. J., Frey, W. H., II, Bordayo, E. Z., Whittum-Hudson, J.

A., and Hudson, A. P., "*Chlamydophila (Chlamydia) pneumoniae* in the Alzheimer's brain," *FEMS Immunology & Medical Microbiology* 2006; 48(3): 355–66. https://doi.org/10.1111/j.1574-695X.2006.00154.x

30. Hung, C.-C., Chang, C.-C., Huang, C.-W., Nouchi, R., and Cheng, C.-H., "Gut microbiota in patients with Alzheimer's disease spectrum: A systematic review and meta-analysis," *Aging* 2022;14(1): 477–96. https://doi.org/10.18632/aging.203826

31. Korf, J. M., Ganesh, B. P., and McCullough, L. D., "Gut dysbiosis and age-related neurological diseases in females," *Neurobiology of Disease* 2022;168: 105695. https://doi.org/10.1016/j.nbd.2022.105695

32. Liu, M., and Zhong, P., "Modulating the gut microbiota as a therapeutic intervention for Alzheimer's disease," *Indian Journal of Microbiology* 2022;62: 494–504. https://doi.org/10.1007/s12088-022-01025-w

33. Haran, J. P., Bhattarai, S. K., Foley, S. E., Dutta, P., Ward, D. V., Bucci, V., and McCormick, B. A., "Alzheimer's disease microbiome is associated with dysregulation of the anti-inflammatory P-glycoprotein Pathway," *mBio* 2019;10(3): e00632–19. https://doi.org/10.1128/mBio.00632-19

34. Zhang, Y., Shen, Y., Liufu, N., Liu, L., Li, W., Shi, Z., Zheng, H., Mei, X., Chen, C. Y., Jiang, Z., Abtahi, S., Dong, Y., Liang, F., Shi, Y., Cheng, L., Yang, G., Kang, J. X., Wilkinson, J., and Xie, Z., "Transmission of Alzheimer's disease-associated microbiota dysbiosis and its impact on cognitive function: Evidence from mice and patients," *Molecular Psychiatry* 2023;28(10): 4421–37. https://doi.org/10.1038/s41380-023-02216-7

35. Harach, T., Marungruang, N., Duthilleul, N., Cheatham, V., McCoy, K. D., Frisoni, G., Neher, J. J., Fåk, F., Jucker, M., Lasser, T., and Bolmont, "Reduction of Abeta amyloid pathology in APPPS1 transgenic mice in the absence of gut microbiota," *Scientific Reports* 2017;7(1): 41802. https://doi.org/10.1038/srep41802

36. Zhang, Y., and Jian, W., "Signal pathways and intestinal flora through trimethylamine N-oxide in Alzheimer's disease," *Current Protein & Peptide Science* 2023;24(9): 721–36. https://doi.org/10.2174/1389203724666230717125406

37. Praveenraj, S. S., Sonali, S., Anand, N., Tousif, H. A., Vichitra, C., Kalyan, M., Kanna, P. V., Chandana, K. A., Shasthara, P., Mahalakshmi, A. M., Yang, J., Pandi-Perumal, S. R., Sakharkar, M. K., and Chidambaram, S. B., "The role of a gut microbial-derived metabolite, trimethylamine N-oxide (TMAO), in neurological disorders," *Molecular Neurobiology* 2022;59(11): 6684–700. https://doi.org/10.1007/s12035-022-02990-5

38. Zhan, X., Stamova, B., Jin, L.-W., DeCarli, C., Phinney, B., and Sharp,

F. R., "Gram-negative bacterial molecules ssociate with Alzheimer disease pathology," *Neurology* 2016;87(22): 2324–32. https://doi.org / 10.1212/WNL.0000000000003391

39. Soo Kim, H., Kim, S., Shin, S. J., Park, Y. H., Nam, Y., won Kim, C., won Lee, K., Kim, S.-M., Jung, I. D., Yang, D. H., Park, Y.-M., and Moon, M., "Gram-negative bacteria and their lipopolysaccharides in Alzheimer's disease: Pathologic roles and therapeutic implications," *Translational Neurodegeneration* 2021;10(49). https://doi.org/10.1186 /s40035-021-00273-y

40. Brown, G. C., "The endotoxin hypothesis of neurodegeneration," *Journal of Neuroinflammation* 2019;16: 180. https://doi.org/10.1186/s12974-019 -1564-7

41. Leblhuber, F., Geisler, S., Steiner, K., Fuchs, D., and Schütz, B., "Elevated fecal calprotectin in patients with Alzheimer's dementia indicates leaky gut," *Journal of Neural Transmission* 2015;122: 1319–22. https://doi .org/10.1007/s00702-015-1381-9

42. Amor, S., Puentes, F., Baker, D., and van der Valk, P., "Inflammation in neurodegenerative diseases," *Immunology* 2010;129(2): 154–69. https:// doi.org/10.1111/j.1365-2567.2009.03225.x

43. Sánchez-Tapia, M., Mimenza-Alvarado, A., Granados-Domínguez, L., Flores-López, A., López-Barradas, A., Ortiz, V., Pérez-Cruz, C., Sánchez-Vidal, H., Hernández-Acosta, J., and Ávila-Funes, J. A., "The gut microbiota–brain axis during aging, mild cognitive impairment and dementia: Role of tau protein, β-amyloid and LPS in serum and curli protein in stool," *Nutrients* 2023;15(4): 932. https://doi.org/10.3390/nu15040932

44. Wang, L.-M., Wu, Q., Kirk, R. A., Horn, K. P., Ebada Salem, A. H., Hoffman, J. M., Yap, J. T., Sonnen, J. A., Towner, R. A., Bozza, F. A., Rodrigues, R. S., and Morton, K. A., "Lipopolysaccharide endotoxemia induces amyloid-β and p-tau formation in the rat brain," *American Journal of Nuclear Medicine and Molecular Imaging* 2018;8(2): 86–99. https://pubmed.ncbi .nlm.nih.gov/29755842/

45. Li, B., He, Y., Ma, J., Huang, P., Du, J., Cao, L., Wang, Y., Xiao, Q., Tang, H., and Chen, S., "Mild cognitive impairment has similar alterations as Alzheimer's disease in gut microbiota," *Alzheimer's & Dementia* 2019;15(10): 1357–66. https://doi.org/10.1016/j.jalz.2019.07.002

46. Emery, D. C., Shoemark, D. K., Batstone, T. E., Waterfall, C. M., Coghill, J. A., Cerajewska, T. L., Davies, M., West, N. X., and Allen, S. J., "16S rRNA next generation sequencing analysis shows bacteria in Alzheimer's post-mortem brain," *Frontiers in Aging Neuroscience* 2017;9: 195. https://doi.org/10.3389/fnagi.2017.00195

47. Zhao, Y., Jaber, V. R., Pogue, A. I., Sharfman, N. M., Taylor, C., and Lukiw, W. J., "Lipopolysaccharides (LPSs) as potent neurotoxic glycolipids in Alzheimer's disease (AD)," *International Journal of Molecular Sciences* 2022;23(20): 12671. https://doi.org/10.3390/ijms232012671

48. Miklossy, J., Kis, A., Radenovic, A., Miller, L., Forro, L., Martins, R., Reiss, K., Darbinian, N., Darekar, P., Mihaly, L., and Khalili, K., "Beta-amyloid deposition and Alzheimer's type changes induced by Borrelia spirochetes," *Neurobiology of Aging* 2006;27(2): 228–36. https://doi.org/10.1016/j.neurobiolaging.2005.01.018

49. Zou, B., Li, J., Ma, R.-X., Cheng, X.-Y., Ma, R.-Y., Zhou, T.-Y., Wu, Z.-Q., Yao Y., and Li, J., "Gut microbiota is an impact factor based on the brain-gut axis to Alzheimer's disease: A systematic review," *Aging and Disease* 2023;14(3): 964–1678. https://doi.org/10.14336/AD.2022.1127

50. Kao, Y.-C., Ho, P.-C., Tu, Y.-K., Jou, I.-M., and Tsai, K.-J., "Lipids and Alzheimer's disease," *International Journal of Molecular Sciences* 2020;21(4): 1505. https://doi.org/10.3390/ijms21041505

51. Ruan, Y., Tang, J., Guo, X., Li, K., and Li, D., "Dietary fat intake and risk of Alzheimer's disease and dementia: A meta-analysis of cohort studies," *Current Alzheimer Research* 2018;15(9): 869–76. https://doi.org/10.2174/1567205015666180427142350

52. Bozelli, J. C., Jr., Azher, S., and Epand, R. M., "Plasmalogens and chronic inflammatory diseases," *Frontiers in Physiology* 2021;12: 730829. https://doi.org/10.3389/fphys.2021.730829

53. Bizeau, J.-B., Albouery, A., Grégoire, S., Buteau, B., Martine, L., Crépin, M., Bron, A. M., Berdeaux, O., Acar, N., Chassaing, B., and Bringer, M.-A., "Dietary inulin supplementation affects specific plasmalogen species in the brain," *Nutrients* 2022;14(15): 3097. https://doi.org/10.3390/nu14153097

54. Yue, H., Qiu, B., Jia, M., Liu, W., Guo, X.-F., Li, N., Xu, Z.-X., Du, F.-L., Xu, T., and Li, D., "Effects of α-linolenic acid intake on blood lipid profiles: A systematic review and meta-analysis of randomized controlled trials," *Critical Reviews in Food Science and Nutrition* 2021; 61(17): 2894–910. https://doi.org/10.1080/10408398.2020.1790496

55. Oh, Y. K., and Song, J., "Important roles of linoleic acid and α-linolenic acid in regulating cognitive impairment and neuropsychiatric issues in metabolic-related dementia," *Life Sciences* 2024;337: 122356. https://doi.org/10.1016/j.lfs.2023.122356

56. Domingues, A., Almeida, S., da Cruz e Silva, E. F., Oliveira, C. R., and Rego, A. C., "Toxicity of β-amyloid in HEK293 cells expressing NR1/

NR2A or NR1/NR2B *N*-methyl-D-aspartate receptor subunit," *Neurochemistry International* 2007;50(6): 872–80. https://doi.org/10.1016/j .neuint.2007.03.001

57. Poletto, A. C., Furuya, D. T., David-Silva, A., Ebersbach-Silva, P., Santos, C. L., Corrêa-Giannella, M. L., Passarelli, M., and Machado, U. F., "Oleic and linoleic fatty acids downregulate *Slc2a4*/GLUT4 expression via NFKB and SREBP1 in skeletal muscle cells," *Molecular and Cellular Endocrinology* 2015;401: 65–72. https://doi.org/10.1016/j.mce .2014.12.001

58. Desale, S. E., and Chinnathambi, S., "α-linolenic acid induces clearance of tau seeds *via* actin-remodeling in microglia," *Molecular Biomedicine* 2021;2(1): 4. https://doi.org/10.1186/s43556-021-00028-1

59. Kaiser, J., "The most common Alzheimer's risk gene may also protect against memory loss," *Science*, October 7, 2021. https://doi.org /10.1126/science.acx9319

60. Patrick, R. P., "Role of phosphatidylcholine-DHA in preventing APOE4-associated Alzheimer's disease," *FASEB Journal* 2019;33(2): 1554–64. https://doi.org/10.1096/fj.201801412R

61. Calder, P. C., "Docosahexaenoic acid," *Annals of Nutrition and Metabolism* 2016;69(Suppl. 1): 8–21. https://doi.org/10.1159/000448262

62. Qin, Y., Havulinna, A. S., Liu, Y., Jousilahti, P., Ritchie, S. C., Tokolyi, A., Sanders, J. G., Valsta, L., Brożyńska, M., Zhu, Q., Tripathi, A., Vázquez-Baeza, Y., Loomba, R., Cheng, S., Jain, M., Niiranen, T., Lahti, L., Knight, R., Salomaa, V., Inouye, M., and Méric, G., "Combined effects of host genetics and diet on human gut microbiota and incident disease in a single population cohort," *Nature Genetics* 2022;54: 134–42. https://doi. org/10.1038/s41588-021-00991-z

63. Grieneisen, L., Dausani, M., Gould, T. J., Björk, J. R., Grenier, J.-C., Yotova, V., Jansen, D., Gottel, N., Gordon, J. B., Learn, N. H., Gesquiere, L. R., Wango, T. L., Mututua, R. S., Warutere, J. K., Siodi, L., Gilbert, J. A., Barreiro, L. B., Alberts, S. C., Tung, J., Archie, E. A., and Blekhman, R., "Gut microbiome heritability is nearly universal but environmentally contingent," *Science* 2021;373(6551): 181–86. https://doi.org/10.1126/ science.aba5483

64. Tran, T.T.T., Corsini, S., Kellingray, L., Hegarty, C., Le Gall, G., Narbad, A., Müller, M., Tejera, N., O'Toole, P. W., Minihane, A.-M., and Vauzour, D., "*APOE* genotype influences the gut microbiome structure and function in humans and mice: Relevance for Alzheimer's disease pathophysiology," *FASEB Journal* 2019;33(7): 8221–31. https://doi .org/10.1096/fj.201900071R

65. Josefson, D., "Alzheimer's disease rarer among Nigerians than among African Americans," *BMJ* 2001;322(7286): 574. https://www.ncbi.nlm.nih.gov/pmc/articles/PMC1174728/

66. Seo, D.-O., O'Donnell, D., Jain, N., Ulrich, J. D., Herz, J., Li, Y., Lemieux, M., Cheng, J., Hu, H., Serrano, J. R., Bao, X., Franke, E., Karlsson, M., Meier, M., Deng, S., Desai, C., Dodiya, H., Lelwala-Guruge, J., Handley, S. A., Kipnis, J., Sisodia, S. S., Gordon, J. I., and Holtzman, D. M., "ApoE isoform– and microbiota-dependent progression of neurodegeneration in a mouse model of tauopathy," *Science* 2023;379(6628): eadd1236. https://doi.org/10.1126/science.add1236

67. Barberger-Gateau, P., Samieri, C., Féart, C., and Plourde, M., "Dietary omega 3 polyunsaturated fatty acids and Alzheimer's disease: Interaction with apolipoprotein E genotype," *Current Alzheimer Research* 2011;8(5): 479–91. https://doi.org/10.2174/156720511796391926

68. Warnecke, T., Schäfer, K.-H., Claus, I., Del Tredici, K., and Jost, W. H., "Gastrointestinal involvement in Parkinson's disease: Pathophysiology, diagnosis, and management," *NPJ Parkinson's Disease* 2022;8(1): 31. https://doi.org/10.1038/s41531-022-00295-x

69. Rolli-Derkinderen, M., Leclair-Visonneau, L., Bourreille, A., Coron, E., Neunlist, M., and Derkinderen, P., "Is Parkinson's disease a chronic low-grade inflammatory bowel disease?" *Journal of Neurology* 2019;267(8): 2207–13. https://doi.org/10.1007/s00415-019-09321-0

70. Amor, S., Puentes, F., Baker, D., and van der Valk, P., "Inflammation in neurodegenerative diseases," *Immunology* 2010;129(2): 154–69. https://doi.org/10.1111/j.1365-2567.2009.03225.x

71. Wallen, Z. D., Demirkan, A., Twa, G., Cohen, G., Dean, M. N., Standaert, D. G., Sampson, T. R., and Payami, H., "Metagenomics of Parkinson's disease implicates the gut microbiome in multiple disease mechanisms," *Nature Communications* 2022;13: 6958. https://doi.org/10.1038/s41467-022-34667-x

72. Huang, B., Chau, S.W.H., Liu, Y., Chan, J.W.Y., Wang, J., Ma, S. L., Zhang, J., Chan, P.K.S., Yeoh, Y. K., Chen, Z., Zhou, L., Wong, S. H., Mok, V.C.T., To, K. F., Lai, H. M., Ng, S., Trenkwalder, C., Chan, F.K.L., and Wing, Y. K., "Gut microbiome dysbiosis across early Parkinson's disease, REM sleep behavior disorder and their first-degree relatives," *Nature Communications* 2023;14(1): 2501. https://doi.org/10.1038/s41467-023-38248-4

73. Narasimhan, H., Ren, C. C., Deshpande, S., and Sylvia, K. E., "Young at gut—turning back the clock with the gut microbiome," *Microorganisms*. 2021;9(3): 555. https://doi.org/10.3390/microorganisms9030555

74. Yang, X., Ai, P., He, X., Mo, C., Zhang, Y., Xu, S., Lai, Y., Qian, Y., and Xiao, Q., "Parkinson's disease is associated with impaired gut–blood barrier for short-chain fatty acids," *Movement Disorders* 2022;37(8): 1634–43. https://doi.org/10.1002/mds.29063

75. Chen, S.-J., Chi, Y.-C., Ho, C.-H., Yang, W.-S., and Lin, C. H., "Plasma lipopolysaccharide-binding protein reflects risk and progression of Parkinson's disease," *Journal of Parkinson's Disease* 2021;11(3): 1129–39. https://doi.org/10.3233/JPD-212574

76. Zhao, Y., Walker, D. I., Lill, C. M., Bloem, B. R., Darweesh, S.K.L., Pinto-Pacheco, B., McNeil, B., Miller, G. W., Heath, A. K., Frissen, M., Petrova, D., Sánchez, M.-J., Chirlaque, M.-D., Guevara, M., Zibetti, M., Panico, S., Middleton, L., Katzke, V., Kaaks, R., Riboli, E., Masala, G., Sieri, S., Zamora-Ros, R., Amiano, P., Jenab, M., Peters, S., and Vermeulen, R., "Lipopolysaccharide-binding protein and future Parkinson's disease risk: A European prospective cohort," *Journal of Neuroinflammation* 2023;20(1): 170. https://doi.org/10.1186/s12974-023-02846-2

77. Kalia, L. V., and Lang, A. E., "Parkinson's disease," *Lancet* 2015; 386(9996): 896–912. https://doi.org/10.1016/S0140-6736(14)61393-3

78. Goedert, M., Spillantini, M. G., Del Tredici, K., and Braak, H., "100 years of Lewy pathology," *Nature Reviews Neurology* 2013;9(1): 13–24. https://doi.org/10.1038/nrneurol.2012.242

79. Ghaisas, S., Maher, J., and Kanthasamy, A., "Gut microbiome in health and disease: Linking the microbiome-gut-brain axis and environmental factors in the pathogenesis of systemic and neurodegenerative diseases," *Pharmacology & Therapeutics* 2016;158: 52–62. https://doi.org/10.1016/j.pharmthera.2015.11.012

80. Bunyoz, A. H., Christensen, R.H.B., Orlovska-Waast, S., Nordentoft, M., Mortensen, P. B., Petersen, L. V., and Benros, M. E., "Vagotomy and the risk of mental disorders: A nationwide population-based study," *Acta Psychiatrica Scandinavica* 2022;145(1):67–78. https://doi.org/10.1111/acps.13343

81. Pan-Montojo, F., Schwarz, M., Winkler, C., Arnhold, M., O'Sullivan, G. A., Pal, A., Said, J., Marsico, G., Verbavatz, J.-M., Rodrigo-Angulo, M., Gille, G., Funk, R.H.W., and Reichmann, H., "Environmental toxins trigger PD-like progression via increased alpha-synuclein release from enteric neurons in mice," *Scientific Reports* 2012;2: 898. https://doi.org/10.1038/srep00898

82. Kim, S., Kwon, S.-H., Kam, T.-I., Panicker, N., Karuppagounder, S. S., Lee, S., Lee, J. H., Kim, W. R., Kook, M., Foss, C. A., Shen, C., Lee, H., Kulkarni, S., Pasricha, P. J., Lee, G., Pomper, M. G., Dawson, V. L.,

Dawson, T. M., and Ko, H. S., "Transneuronal propagation of pathologic α-synuclein from the gut to the brain models Parkinson's disease," *Neuron* 2019;103(4): 627–41.e7. https://doi.org/10.1016/j.neuron.2019.05.035

83. Wakade, C., Chong, R., Bradley, E., Thomas, B., and Morgan, J., "Up-regulation of GPR109A in Parkinson's disease," *PLoS ONE* 2014;9(10): e109818. https://doi.org/10.1371/journal.pone.0109818

84. Bender, D. A., Earl, C. J., and Lees, A. J., "Niacin depletion in Parkinsonian patients treated with L-dopa, benserazide and carbidopa," *Clinical Science* (London) 1979;56(1): 89–93. https://doi.org/10.1042/cs0560089

85. Karunaratne, T. B., Okereke, C., Seamon, M., Purohit, S., Wakade, C., and Sharma, A., "Niacin and butyrate: Nutraceuticals targeting dysbiosis and intestinal permeability in Parkinson's disease," *Nutrients* 2020;13(1): 28. https://doi.org/10.3390/nu13010028

86. Sun, H., Zhao, F., Liu, Y., Ma, T., Jin, H., Quan, K., Leng, B., Zhao, J., Yuan, X., Li, Z., Li, F., Kwok, L.-Y., Zhang, S., Sun, Z., Zhang, J., and Zhang, H., "Probiotics synergized with conventional regimen in managing Parkinson's disease," *npj Parkinson's Disease* 2022;8(1): 62. https://doi.org/10.1038/s41531-022-00327-6

CHAPTER 9: THE DOS AND DON'TS OF THE GUT-BRAIN PARADOX PROGRAM(S)

1. Corbin, K. D., Carnero, E. A., Dirks, B., Igudesman, D., Yi, F., Marcus, A., Davis, T. L., Pratley, R. E., Rittmann, B. E., Krajmalnik-Brown, R., and Smith, S. R., "Host-diet-gut microbiome interactions influence human energy balance: A randomized clinical trial," *Nature Communications* 2023;14: 3161. https://doi.org/10.1038/s41467-023-38778-x

2. Wastyk, H. C., Fragiadakis, G. K., Perelman, D., Dahan, D., Merrill, B. D., Yu, F. B., Topf, M., Gonzalez, C. G., Van Treuren, W., Han, S., Robinson, J. L., Elias, J. E., Sonnenburg, E. D., Gardner, C. D., and Sonnenburg, J. L., "Gut-microbiota-targeted diets modulate human immune status," *Cell* 2021;184(16): 4137–53.e14. https://doi.org/10.1016/j.cell.2021.06.019

3. Balasubramanian, R., Schneider, E., Gunnigle, E., Cotter, P. D., and Cryan, J. F., "Fermented foods: Harnessing their potential to modulate the microbiota-gut-brain axis for mental health," *Neuroscience & Biobehavioral Reviews* 2024;158: 105562. https://doi.org/10.1016/j.neubiorev.2024.105562

4. Duncan, S. H., Louis, P., and Flint, H. J., "Lactate-utilizing bacteria, isolated from human feces, that produce butyrate as a major fermentation

product," *Applied and Environmental Microbiology* 2004;70(10): 5810–17. https://doi.org/10.1128/AEM.70.10.5810-5817.2004

5. Pirinen, E., Kuulasmaa, T., Pietilä, M., Heikkinen, S., Tusa, M., Itkonen, P., Boman, S., Skommer, J., Virkamäki, A., Hohtola, E., Kettunen, M., Fatrai, S., Kansanen, E., Koota, S., Niiranen, K., Parkkinen, J., Levonen, A.-L., Ylä-Herttuala, S., Hiltunen, J. K., Alhonen, L., Smith, U., and Jänne, J., "Enhanced polyamine catabolism alters homeostatic control of white adipose tissue mass, energy expenditure, and glucose metabolism," *Molecular and Cellular Biology* 2007;27 (13): 4953–67. https://mcb.asm.org/content/27/13/4953

6. Parlindungan, E., Lugli, G. A., Ventura, M., van Sinderen, D., and Mahony, J., "Lactic acid bacteria diversity and characterization of probiotic candidates in fermented meats," *Foods* 2021;10(7): 1519. https://doi.org/10.3390/foods10071519

7. Cortés-Martín, A., Selma, M. V., Tomás-Barberán, F. A., González-Sarrías, A., and Espín, J. C., "Where to look into the puzzle of polyphenols and health? The postbiotics and gut microbiota associated with human metabotypes," *Molecular Nutrition & Food Research* 2020;64(9): e1900952. https://doi.org/10.1002/mnfr.201900952

8. Annunziata, G., Maisto, M., Schisano, C., Ciampaglia, R., Narciso, V., Tenore, G. C., and Novellino, E., "Effects of grape pomace polyphenolic extract (Taurisolo®) in reducing TMAO serum levels in humans: Preliminary results from a randomized, placebo-controlled, cross-over study," *Nutrients* 2019;11(1): 139. https://doi.org/10.3390/nu11010139

9. Brand, M. D., "Uncoupling to survive? The role of mitochondrial inefficiency in ageing," *Experimental Gerontology* 2000;35(6–7): 811–20. https://doi.org/10.1016/s0531-5565(00)00135-2

10. Bernardi, S., Del Bo', C., Marino, M., Gargari, G., Cherubini, A., Andrés-Lacueva, C., Hidalgo-Liberona, N., Peron, G., González-Dominguez, R., Kroon, P., Kirkup, B., Porrini, M., Guglielmetti, S., and Riso, P., "Polyphenols and intestinal permeability: Rationale and future perspectives," *Journal of Agricultural and Food Chemistry* 2020;68(7): 1816–29. https://doi.org/10.1021/acs.jafc.9b02283

11. Chandrasekaran, K., Salimian, M., Konduru, S. R., Choi, J., Kumar, P., Long, A., Klimova, N., Ho, C.-Y., Kristian, T., and Russell, J. W., "Overexpression of Sirtuin 1 protein in neurons prevents and reverses experimental diabetic neuropathy," *Brain* 2019;142(12): 3737–52. https://doi.org/10.1093/brain/awz324

12. Turiaco, F., Cullotta, C., Mannino, F., Bruno, A., Squadrito, F., Pallio, G., and Irrera, N., "Attention deficit hyperactivity disorder (ADHD) and

polyphenols: A systematic review," *International Journal of Molecular Sciences* 2024;25(3): 1536. https://doi.org/10.3390/ijms25031536

13. Ma, X., Sun., Z., Han, X., Li, S., Jiang, X., Chen, S., Zhang, J., and Lu, H., "Neuroprotective effect of resveratrol via activation of Sirt1 signaling in a rat model of combined diabetes and Alzheimer's disease," *Frontiers in Neuroscience* 2019;13: 1400. https://doi.org/10.3389/fnins.2019.01400

14. Dasgupta, B., and Milbrandt, J., "Resveratrol stimulates AMP kinase activity in neurons," *Proceedings of the National Academy of Sciences USA* 2007;104(17): 7217–22. https://doi.org/10.1073/pnas.0610068104

15. Ryu, D., Mouchiroud, L., Andreux, P. A., Katsyuba, E., Moullan, N., Nicolet-dit-Félix, A. A., Williams, E. G., Jha, P., Lo Sasso, G., Huzard, D., Aebischer, P., Sandi, C., Rinsch, C., and Auwerx, J., "Urolithin A induces mitophagy and prolongs lifespan in *C. elegans* and increases muscle function in rodents," *Nature Medicine* 2016;22(8): 879–88. https://doi.org/10.1038/nm.4132

16. Selma, M. V., Beltrán, D., Luna, M. C., Romo-Vaquero, M., García-Villalba, R., Mira, A., Espín J. C., and Tomás-Barberán, F. A., "Isolation of human intestinal bacteria capable of producing the bioactive metabolite isourolithin A from ellagic acid," *Frontiers in Microbiology* 2017;8: 1521. https://doi.org/10.3389/fmicb.2017.01521

17. Aktar, S., Ferdousi, F., Kondo, S., and Isoad, H., "Transcriptomics and biochemical evidence of trigonelline ameliorating learning and memory decline in the senescence-accelerated mouse prone 8 (SAMP8) model by suppressing proinflammatory cytokines and elevating neurotransmitter release," *GeroScience* 2024;46: 1671–91. https://doi.org/10.1007/s11357-023-00919-x

18. Li, Z., Teng, J., Lyu, Y., Hu, X., Zhao, Y., and Wang, M., "Enhanced antioxidant activity for apple juice fermented with *Lactobacillus plantarum* ATCC14917," *Molecules* 2018;24(1): 51. https://doi.org/10.3390/molecules24010051

19. Sharma, R., Diwan, B., Singh, B. P., and Kulshrestha, S., "Probiotic fermentation of polyphenols: Potential sources of novel functional foods," *Food Production, Processing and Nutrition* 2022;4: 21. https://doi.org/10.1186/s43014-022-00101-4

20. Pérez-Jiménez, J., Neveu, V., Vos, F., and Scalbert, A., "Identification of the 100 richest dietary sources of polyphenols: An application of the Phenol-Explorer database," *European Journal of Clinical Nutrition* 2010;64: S112–20. https://doi.org/10.1038/ejcn.2010.221

21. Dong, J., Ye, F., Lin, J., He, H., and Song, Z., "The metabolism and function of phospholipids in mitochondria," *Mitochondrial Communications* 2023;1: 2–12. https://doi.org/10.1016/j.mitoco.2022.10.002

22. Li, Y., Lai, W., Zheng, C., Babu, J. R., Xue, C., Ai, Q., and Huggins, K. W., "Neuroprotective effect of stearidonic acid on amyloid β-induced neurotoxicity in rat hippocampal cells," *Antioxidants* (Basel) 2022;11(12): 2357. https://doi.org/10.3390/antiox11122357

23. Zhu, X., Wang, B., Zhang, X., Chen, X., Zhu, J., Zou, Y., and Li, J., "Alpha-linolenic acid protects against lipopolysaccharide-induced acute lung injury through anti-inflammatory and anti-oxidative path-ways," *Microbial Pathogenesis* 2020;142: 104077. https://doi.org/10.1016/j.micpath.2020.104077

24. Shen, J., Liu, Y., Wang, X., Bai, J., Lin, L., Luo, F., and Zhong, H., "A comprehensive review of health-benefiting components in rapeseed oil," *Nutrients* 2023;15(4): 999. https://doi.org/10.3390/nu15040999

25. Todorov, H., Kollar, B., Bayer, F., Brandão, I., Mann, A., Mohr, J., Pontarollo, G., Formes, H., Stauber, R., Kittner, J. M., Endres, K., Watzer, B., Nockher, W. A., Sommer, F., Gerber, S., and Reinhardt, C., "α-linolenic acid-rich diet influences microbiota composition and villus morphology of the mouse small intestine," *Nutrients* 2020;12(3): 732. https://doi.org/10.3390/nu12030732

26. Baxheinrich, A., Stratmann, B., Lee-Barkey, Y. H., Tschoepe, D., and Wahrburg, U., "Effects of a rapeseed oil-enriched hypoenergetic diet with a high content of α-linolenic acid on body weight and cardiovascular risk profile in patients with the metabolic syndrome," *British Journal of Nutrition* 2012;108(4): 682–91. https://doi.org/10.1017/S0007114512002875

27. Gao, X., Chang, S., Liu, S., Peng, L., Xie, J., Dong, W., Tian, Y., and Sheng, J., "Correlations between α-linolenic acid-improved multitissue homeostasis and gut microbiota in mice fed a high-fat diet," *mSystems* 2020;5(6): e00391–20. https://doi.org/10.1128/mSystems.00391-20

28. Kawamura, A., Nemoto, K., and Sugita, M., "Effect of 8-week intake of the *n*-3 fatty acid-rich perilla oil on the gut function and as a fuel source for female athletes: A randomised trial," *British Journal of Nutrition* 2022;129(6): 981–91. https://doi.org/10.1017/S0007114522001805

29. Kawashima, H., "Intake of arachidonic acid-containing lipids in adult humans: Dietary surveys and clinical trials," *Lipids in Health and Disease* 2019;18(101). https://doi.org/10.1186/s12944-019-1039-y

30. Depauw, S., Bosch, G., Hesta, M., Whitehouse-Tedd, K., Hendriks, W. H., Kaandorp, J., and Janssens, G.P.J., "Fermentation of animal components in strict carnivores: A comparative study with cheetah f ecal inoculum," *Journal of Animal Science* 2012;90(8): 2540–48. https://doi.org/10.2527/jas.2011-4377

31. Mark, K. A., Dumas, K. J., Bhaumik, D., Schilling, B., Davis, S., Oron, T. R., Sorensen, D. J., Lucanic, M., Brem, R. B., Melov, S., Ramanathan, A., Gibson, B. W., and Lithgow, G. J., "Vitamin D promotes protein homeostasis and longevity via the stress response pathway genes *skn-1, ire-1*, and *xbp-1*," *Cell Reports* 2016;17(5): 1227–37. https://doi.org/10.1016/j.celrep.2016.09.086

32. Thomas, R. L., Jiang, L., Adams, J. S., Xu, Z. Z., Shen, J., Janssen, S., Ackermann, G., Vanderschueren, D., Pauwels, S., Knight, R., Orwoll, E. S., and Kado, D. M., "Vitamin D metabolites and the gut microbiome in older men," *Nature Communications* 2020;11: 5997. https://doi.org/10.1038/s41467-020-19793-8

33. Singh, P., Rawat, A., Alwakeel, M., Sharif, E., and Al Khodor, S., "The potential role of vitamin D supplementation as a gut microbiota modifier in healthy individuals," *Scientific Reports* 2020;10(1): 21641. https://doi.org/10.1038/s41598-020-77806-4

34. Hahn, J., Cook, N. R., Alexander, E. K., Friedman, S., Walter, J., Bubes, V., Kotler, G., Lee, I.-M., Manson, J. E., and Costenbader, K. H., "Vitamin D and marine omega 3 fatty acid supplementation and incident autoimmune disease: VITAL randomized controlled trial," *BMJ* 2022;376: e066452. https://doi.org/10.1136/bmj-2021-066452

35. Ghahremani, M., Smith, E. E., Chen, H.-Y., Creese, B., Goodarzi, Z., and Ismail, Z., "Vitamin D supplementation and incident dementia: Effects of sex, *APOE*, and baseline cognitive status," *Alzheimer's & Dementia* 2023;15(1): e12404. https://doi.org/10.1002/dad2.12404

36. Garland, C. F., French, C. B., Baggerly, L. L., and Heaney, R. P., "Vitamin D supplement doses and serum 25-hydroxyvitamin D in the range associated with cancer prevention," *Anticancer Research* 2011;31(2): 607–11. https://pubmed.ncbi.nlm.nih.gov/21378345/

37. Xu, C., Zhang, J., Mihai, D. M., and Washington, I., "Light-harvesting chlorophyll pigments enable mammalian mitochondria to capture photonic energy and produce ATP," *Journal of Cell Science* 2014;127(2): 388–99. https://doi.org/10.1242/jcs.134262

38. Deopurkar, R., Ghanim, H., Friedman, J., Abuaysheh, S., Sia, C. L., Mohanty, P., Viswanathan, P., Chaudhuri, A., and Dandona, P., "Differential effects of cream, glucose, and orange juice on inflammation, endotoxin, and the expression of toll-like receptor-4 and suppressor of cytokine signaling-3," *Diabetes Care* 2010;33(5): 991–97. https://doi.org/10.2337/dc09-1630

39. Anggraini, H., Tongkhao, K., and Chanput, W., "Reducing milk allergenicity of cow, buffalo, and goat milk using lactic acid bacteria fermen-

tation," *AIP Conference Proceedings* 2018;2021(1): 070010. https://doi .org/10.1063/1.5062808

40. Hasegawa, Y., Pei, R., Raghuvanshi, R., Liu, Z., and Bolling, B. W., "Yogurt supplementation attenuates insulin resistance in obese mice by reducing metabolic endotoxemia and inflammation," *Journal of Nutrition* 2023;153(3): 703–12. https://doi.org/10.1016/j.tjnut .2023.01.021

41. Pei, R., DiMarco, D. M., Putt, K. K., Martin, D. A., Chitchumroon-chokchai, C., Bruno, R. S., and Bolling, B. W., "Premeal low-fat yogurt consumption reduces postprandial inflammation and markers of en-dotoxin exposure in healthy premenopausal women in a randomized controlled trial," *Journal of Nutrition* 2018;148(6): 910–16. https://doi .org/10.1093/jn/nxy046. Erratum in: *Journal of Nutrition* 2018;148(10): 1698. https://doi.org/10.1093/jn/nxy178

42. Nieddu, A., Vindas, L., Errigo, A., Vindas, J., Pes, G. M., and Dore, M. P., "Dietary habits, anthropometric features and daily performance in two independent long-lived populations from *Nicoya peninsula* (Costa Rica) and *Ogliastra* (Sardinia)," *Nutrients* 2020;12(6): 1621. https://doi. org/10.3390/nu12061621

43. Bell, C., "Six fascinating facts about sheep milk," *Shepherds Purse* (blog), July 16, 2020. https://blog.shepherdspurse.co.uk/blog/six -facts-benefits-sheep-milk

44. Li, H., Li, S., Yang, H., Zhang, Y., Zhang, S., Ma, Y., Hou, Y., Zhang, X., Niu, K., Borné, Y., and Wang, Y., "Association of ultrapro-cessed food consumption with risk of dementia: A prospective co-hort study," *Neurology* 2022;99(10): e1056–66. https://doi.org/10.1212 /WNL.0000000000200871

45. Peh, M. T., Anwar, A. B., Ng, D. S., Atan, M. S., Kumar, S. D., and Moore, P. K., "Effect of feeding a high fat diet on hydrogen sulfide (H_2S) metab-olism in the mouse," *Nitric Oxide: Biology and Chemistry* 2014;41: 138–45. https://doi.org/10.1016/j.niox.2014.03.002

46. Gabriela, P., Tan, J., Bartlomiej, J., Kaakoush, N. O., Angelatos, A. S., O'Sullivan, J., Koay, Y. C., Sierro, F., Davis, J., Divakarla, S. K., Khanal, D., Moore, R. J., Stanley, D., Chrzanowski, W., and Macia, L., "Im-pact of food additive titanium dioxide (E 171) on gut microbiota-host interaction," *Frontiers in Nutrition* 2019;6: 57. https://doi.org/10.3389 /fnut.2019.00057

47. Holton, K. F., Ndege, P. K., and Clauw, D. J., "Dietary correlates of chronic widespread pain in Meru, Kenya," *Nutrition* 2018;53: 14–19. https://doi .org/10.1016/j.nut.2018.01.016

48. Bocos, C., "Fructose consumption hampers gasotransmitter production," *Academia Letters* 2021: n. pag. Web.

49. Pase, M. P., Himali, J. J., Jacques, P. F., DeCarli, C., Satizabal, C. L., Aparicio, H., Vasan, R. S., Beiser, A. S., and Deshadri, S., "Sugary beverage intake and preclinical Alzheimer's disease in the community," *Alzheimer's & Dementia* 2017;13(9): 955–64. https://doi.org/10.1016/j.jalz.2017.01.024

50. Chong, C. P., Shahar, S., Haron, H., and Din, N. C., "Habitual sugar intake and cognitive impairment among multi-ethnic Malaysian older adults," *Clinical Interventions in Aging* 2019;14: 1331–42. https://doi.org/10.2147/cia.s211534

51. Cohen, J.F.W., Rifas-Shiman, S. L., Young, J., and Oken, E., "Associations of prenatal and child sugar intake with child cognition," *American Journal of Preventive Medicine* 2018;54(6): 727–35. https://doi.org/10.1016/j.amepre.2018.02.020

52. De Punder, K., and Pruimboom, L., "The dietary intake of wheat and other cereal grains and their role in inflammation," *Nutrients* 2013;5(3): 771–87. https://doi.org/10.3390/nu5030771

53. Schumacher, U., von Armansperg, N. G., Kreipe, H., and Welsch, U., "Lectin binding and uptake in human (myelo)monocytic cell lines: HL60 and U937," *Ultrastructural Pathology* 1996;20(5): 463–471. https://doi.org/10.3109/01913129609016350

54. Kitada, M., Ogura, Y., Monno, I., and Koya, D., "The impact of dietary protein intake on longevity and metabolic health," *eBioMedicine* 2019;43: 632–40. https://doi.org/10.1016/j.ebiom.2019.04.005

55. Fontana, L., Weiss, E. P., Villareal, D. T., Klein, S., and Holloszy, J. O., "Long-term effects of calorie or protein restriction on serum IGF-1 and IGFBP-3 concentration in humans," *Aging Cell* 2008;7(5): 681–87. https://doi.org/10.1111/j.1474-9726.2008.00417.x

56. Sheikhi, A., Siassi, F., Djazayery, A., Guilani, B., and Azadbakht, L., "Plant and animal protein intake and its association with depression, anxiety, and stress among Iranian women," *BMC Public Health* 2023;23(1): 161. https://doi.org/10.1186/s12889-023-15100-4

57. Mayneris-Perxachs, J., Castells-Nobau, A., Arnoriaga-Rodríguez, M., Martin, M., de la Vega-Correa, L., Zapata, C., Burokas, A., Blasco, G., Coll, C., Escrichs, A., Biarnés, C., Moreno-Navarrete, J. M., Puig, J., Garre-Olmo, J., Ramos, R., Pedraza, S., Brugada, R., Vilanova, J. C., Serena, J., Gich, J., and Fernández-Real, J., "Microbiota alterations in proline metabolism impact depression," *Cell Metabolism* 2022;34(5): 681–701.

e10. https://doi.org/10.1016/j.cmet.2022.04.001

58. Yang, L., Shen, J., Liu, C., Kuang, Z., Tang, Y., Qian, Z., Guan, M., Yang, Y., Zhan, Y., Li, N., and Li, X., "Nicotine rebalances NAD+ homeostasis and improves aging-related symptoms in male mice by enhancing NAMPT activity," *Nature Communications* 2023;14(1): 900. https://doi.org/10.1038/s41467-023-36543-8

59. Evans, S., Gray, M. A., Dowell, N. G., Tabet, N., Tofts, P. S., King, S. L., and Rusted, J. M., "APOE E4 carriers show prospective memory enhancement under nicotine, and evidence for specialisation within medial BA10," *Neuropsychopharmacology* 2013;38(4): 655–63. https://doi.org/10.1038/npp.2012.230

60. Terpinskaya, T. I., Osipov, A. V., Kryukova, E. V., Kudryavtsev, D. S., Kopylova, N. V., Yanchanka, T. L., Palukoshka, A. F., Gondarenko, E. A., Zhmak, M. N., Tsetlin, V. I., and Utkin, Y. N., "α-conotoxins and α-cobratoxin promote, while lipoxygenase and cyclooxygenase inhibitors suppress the proliferation of glioma C6 cells," *Marine Drugs* 2021;19(2): 118. https://doi.org/10.3390/md19020118

61. Waterhouse, U., Brennan, K. A., and Ellenbroek, B. A., "Nicotine self-administration reverses cognitive deficits in a rat model for schizophrenia," *Addiction Biology* 2018;23(2): 620–30. https://doi.org/10.1111/adb.12517

62. Van Schalkwyk, G. I., Lewis, A. S., Qayyum, Z., Koslosky, K., Picciotto, M. R., and Volkmar, F. R., "Reduction of aggressive episodes after repeated transdermal nicotine administration in a hospitalized adolescent with autism spectrum disorder," *Journal of Autism and Developmental Disorders* 2015;45(9): 3061–66. https://doi.org/10.1007/s10803-015-2471-0

63. Yoshida, T., Sakane, N., Umekawa, T., Kogure, A., Kondo, M., Kumamoto, K., Kawada, T., Nagase, I., and Saito, M., "Nicotine induces uncoupling protein 1 in white adipose tissue of obese mice," *International Journal of Obesity and Related Metabolic Disorders* 1999;23(6): 570–75. https://doi.org/10.1038/sj.ijo.0800870

64. Mappin-Kasirer, B., Pan, H., Lewington, S., Kizza, J., Gray, R., Clarke, R., and Peto, R., "Tobacco smoking and the risk of Parkinson disease: A 65 year follow up of 30,000 male British doctors," *Neurology* 2020;94(20): e2132–38. https://doi.org/10.1212/WNL.0000000000009437

65. van Duijn, C. M., and Hofman, A., "Relation between nicotine intake and Alzheimer's disease," *BMJ* 1991;302(6791): 1491–94. https://doi.org/10.1136/bmj.302.6791.1491

66. Chen, Y., Zhao, M., Ji, K., Li, J., Wang, S., Lu, L., Chen, Z., and Zeng, J., "Association of nicotine dependence and gut microbiota: A bidirectional two-sample Mendelian randomization study," *Frontiers in Immunology* 2023;14: 1244272. https://doi.org/10.3389/fimmu.2023.1244272

CHAPTER 10: WHEN AND WHAT TO EAT ON THE GUT-BRAIN PARADOX PROGRAM(S)

1. Yihang, Z., Jia, M., Chen, W., and Liu, Z., "The neuroprotective effects of intermittent fasting on brain aging and neurodegenerative diseases via regulating mitochondrial function," *Free Radical Biology and Medicine* 2022;182: 206–18. https://doi.org/10.1016/j.freeradbiomed.2022.02.021
2. Starr, M. E., Steele, A. M., Cohen, D. A., and Saito, H., "Short-term dietary restriction rescues mice from lethal abdominal sepsis and endotoxemia and reduces the inflammatory/coagulant potential of adipose tissue," *Critical Care Medicine* 2016;44(7): e509–19. https://doi.org/10.1097/CCM.0000000000001475
3. Jantsch, J., da Silva Rodrigues, F., de Farias Fraga, G., Eller, S., Silveira, A. K., Moreira, J.C.F., Giovenardi, M., and Guedes, R. P., "Calorie restriction mitigates metabolic, behavioral and neurochemical effects of cafeteria diet in aged male rats," *Journal of Nutritional Biochemistry* 2023;119: 109371. https://doi.org/10.1016/j.jnutbio.2023.109371
4. Brandhorst, S., Levine, M. E., Wei, M., Shelehchi, M., Morgan, T. E., Nayak, K. S., Dorff, T., Hong, K., Crimmins, E. M., Cohen, P., and Longo, V. D., "Fasting-mimicking diet causes hepatic and blood markers changes indicating reduced biological age and disease risk," *Nature Communications* 2024;15: 1309. https://doi.org/10.1038/s41467-024-45260-9
5. Erlanson-Albertsson, C., and Stenkula, K. G., "The importance of food for endotoxemia and an inflammatory response," *International Journal of Molecular Sciences* 2021;22(17): 9562. https://doi.org/10.3390/ijms22179562

CHAPTER 11: THE CHICKEN AND THE SEA (THE GUT-BRAIN PARADOX MODIFIED CARNIVORE DIET)

1. Gershenzon, J., and Ullah, C., "Plants protect themselves from herbivores by optimizing the distribution of chemical defenses," *Proceedings of the National Academy of Sciences USA* 2022;119(4): e2120277119. https://doi.org/10.1073/pnas.2120277119

2. Karamad, D., Khosravi-Darani, K., Khaneghah, A. M., and Miller, A. W., "Probiotic oxalate-degrading bacteria: New insight of environmental variables and expression of the oxc and frc genes on oxalate degradation activity," *Foods* 2022;11(18): 2876. https://doi.org/10.3390/foods11182876

3. Davies, L. R., and Varki, A., "Why is *N*-glycolylneuraminic acid rare in the vertebrate brain?" *SialoGlyco Chemistry and Biology I* 2015;366: 31–54. https://doi.org/10.1007/128_2013_419

4. Rebelo, A. L., Chevalier, M. T., Russo, L., and Pandit, A., "Role and therapeutic implications of protein glycosylation in neuroinflammation," 2022;28(4): 270–89. https://doi.org/10.1016/j.molmed.2022.01.004

5. Rosenbaum, M., Hall, K. D., Guo, J., Ravussin, E., Mayer, L. S., Reitman, M. L., Smith, S. R., Walsh, B. T., and Leibel, R. L., "Glucose and lipid homeostasis and inflammation in humans following an isocaloric ketogenic diet," *Obesity* 2019;27(6): 971–81. https://doi.org/10.1002/oby.22468

6. Andrews, Z. B., Diano, S., and Horvath, T. L., "Mitochondrial uncoupling proteins in the CNS: In support of function and survival," *Nature Reviews Neuroscience* 2005;6(11): 829–40. https://doi.org/10.1038/nrn1767

7. Morris, G., Puri, B. K., Maes, M., Olive, L., Berk, M., and Carvalho, A. F., "The role of microglia in neuroprogressive disorders: Mechanisms and possible neurotherapeutic effects of induced ketosis," *Progress in Neuro-Psychopharmacology and Biological Psychiatry* 2020;99: 109858. https://doi.org/10.1016/j.pnpbp.2020.109858

8. Smolensky, I. V., Zajac-Bakri, K., Gass, P., and Inta, D., "Ketogenic diet for mood disorders from animal models to clinical application," *Journal of Neural Transmission* (Vienna) 2023;130(9): 1195–205. https://doi.org/10.1007/s00702-023-02620-x

9. Smith, E. A., and Macfarlane, G. T., "Enumeration of amino acid fermenting bacteria in the human large intestine: Effects of pH and starch on peptide metabolism and dissimilation of amino acids," *FEMS Microbiology Ecology* 1998;25(4): 355–68. https://doi.org/10.1016/S0168-6496(98)00004-X

10. Sethi, S., Wakeham, D., Ketter, T., Hooshmand, F., Bjornstad, J., Richards, B., Westman, E., Krauss, R. M., and Saslow, L., "Ketogenic diet intervention on metabolic and psychiatric health in bipolar and schizophrenia: A pilot trial," *Psychiatry Research* 2024;335: 115866. https://doi.org/10.1016/j.psychres.2024.115866

11. Danan, A., Westman, E. C., Saslow, L. R., and Ede, G., "The ketogenic

diet for refractory mental illness: A retrospective analysis of 31 inpatients," *Frontiers in Psychiatry* 2022;13: 951376. https://doi.org/10.3389/fpsyt.2022.951376

12. Murray, E. R., Kemp, M., and Nguyen, T. T., "The microbiota-gut-brain axis in Alzheimer's disease: A review of taxonomic alterations and potential avenues for interventions," *Archives of Clinical Neuropsychology* 2022;37(3): 595–607. https://doi.org/10.1093/arclin/acac008

13. Garner, S., Davies, E., Barkus, E., and Kraeuter, A.-K., "Ketogenic diet has a positive association with mental and emotional well-being in the general population," *Nutrition* 2024;124: 112420. https://doi.org/10.1016/j.nut.2024.112420

14. Kong, D., Sun, J.-x., Yang, J.-q., Li, Y.-s., Bi, K., Zhang, Z.-y., Wang K.-h., Luo, H.-y., Zhu, M., and Xu, Y., "Ketogenic diet: A potential adjunctive treatment for substance use disorders," *Frontiers in Nutrition* 2023;10: 1191903. https://doi.org/10.3389/fnut.2023.1191903

15. Cândido, T.L.N., da Silva, L. E., Tavares, J. F., Conti, A.C.M., Rizzardo, R.A.G., and Gonçalves Alfenas, R. C., "Effects of dietary fat quality on metabolic endotoxaemia: A systematic review," *British Journal of Nutrition* 2020;124(7): 654–67. https://doi.org/10.1017/S0007114520001658

16. Bashir, S., Fezeu, L. K., Leviatan Ben-Arye, S., Yehuda, S., Reuven, E. M., Szabo de Edelenyi, F., Fellah-Hebia, I., Le Tourneau, T., Imbert-Marcille, B. M., Drouet, E. B., Touvier, M., Roussel, J.-C., Yu, H., Chen, X., Hercberg, S., Cozzi, E., Soulillou, J.-P., Galan, P., and Padler-Karavani, V., "Association between Neu5Gc carbohydrate and serum antibodies against it provides the molecular link to cancer: French NutriNet-Santé study," *BMC Medicine* 2020;18(1): 262. https://doi.org/10.1186/s12916-020-01721-8

17. Davies, L.R.L., and Varki, A., "Why is *N*-glycolylneuraminic acid rare in the vertebrate brain?" *SialoGlyco Chemistry and Biology I* 2015;366: 31–54. https://doi.org/10.1007/128_2013_419

18. Banda, K., Gregg, C. J., Chow, R., Varki, N. M., and Varki, A., "Metabolism of vertebrate amino sugars with *N*-glycolyl groups: Mechanisms underlying gastrointestinal incorporation of the non-human sialic acid xeno-autoantigen *N*-glycolylneuraminic acid," *Journal of Biological Chemistry* 2012;287(34): 28852–64. https://doi.org/10.1074/jbc.M112.364182

19. Boligan, K. F., Oechtering, J., Keller, C. W., Peschke, B., Rieben, R., Bovin, N., Kappos, L., Cummings, R. D., Kuhle, J., von Gunten, S., and Lünemann, J. D., "Xenogeneic Neu5Gc and self-glycan Neu5Ac

epitopes are potential immune targets in MS," *Neurology Neuroimmunology & Neuroinflammation* 2020;7(2): e676. https://doi.org/10.1212/NXI.0000000000000676

20. Kawanishi, K., Coker, J. K., Grunddal, K. V., Dhar, C., Hsiao, J., Zengler, K., Varki, N., Varki, A., and Gordts, P.L.S.M., "Dietary Neu5Ac intervention protects against atherosclerosis associated with human-like Neu5Gc loss—brief report," *Arteriosclerosis, Thrombosis, and Vascular Biology* 2021;41(11): 2730–39. https://doi.org/10.1161/ATVBAHA.120.315280

21. Lin, X., Yao, H., Guo, J., Huang, Y., Wang, W., Yin, B., Li, X., Wang, T., Li, C., Xu, X., Zhou, G., Voglmeir, J., and Liu, L., "Protein glycosylation and gut microbiota utilization can limit the in vitro and in vivo metabolic cellular incorporation of Neu5Gc," *Molecular Nutrition & Food Research* 2022;66(5): e2100615. https://doi.org/10.1002/mnfr.202100615

22. Biemans, Y., Bach, D., Behrouzi, P., Horvath, S., Kramer, C. S., Liu, S., Manson, J. E., Shadyab, A. H., Stewart, J., Whitsel, E. A., Yang, B., de Groot, L., and Grootswagers, P., "Identifying the relation between food groups and biological ageing: A data-driven approach," *Age and Ageing* 2024;53(Suppl. 2): ii20–29. https://doi.org/10.1093/ageing/afae038

23. Schroeder, S., Hofer, S. J., Zimmermann, A., Pechlaner, R., Dammbrueck, C., Pendl, T., Marcello, G. M., Pogatschnigg, V., Bergmann, M., Müller, M., Gschiel, V., Ristic, S., Tadic, J., Iwata, K., Richter, G., Farzi, A., Üçal, M., Schäfer, U., Poglitsch, M., Royer, P., Mekis, R., Agreiter, M., Tölle, R. C., Sótonyi, P., Willeit, J., Mairhofer, B., Niederkofler, H., Pallhuber, I., Rungger, G., Tilg, H., Defrancesco, M., Marksteiner, J., Sinner, F., Magnes, C., Pieber, T. R., Holzer, P., Kroemer, G., Carmona-Gutierrez, D., Scorrano, L., Dengjel, J., Madl, T., Sedej, S., Sigrist, S. J., Rácz, B., Kiechl, S., Eisenberg, T., and Madeo, F., "Dietary spermidine improves cognitive function," *Cell Reports* 2021;35(2): 108985. https://doi.org/10.1016/j.celrep.2021.108985

24. Milne, M. H., De Frond, H., Rochman, C. M., Mallos, N. J., Leonard, G. H., and Baechler, B. R., "Exposure of U.S. adults to microplastics from commonly-consumed proteins," *Environmental Pollution* 2024;343: 123233. https://doi.org/10.1016/j.envpol.2023.123233

25. Hui, M., Jia, X., Li, X., Lazcano-Silveira, R., and Shi, M., "Anti-inflammatory and antioxidant effects of liposoluble C60 at the cellular, molecular, and whole-animal levels," *Journal of Inflammation Research* 2023;16: 83–93. https://doi.org/10.2147/JIR.S386381

26. de Oliveira Otto, M. C., Nettleton, J. A., Lemaitre, R. N., Steffen,

L. M., Kromhout, D., Rich, S. S., Tsai, M. Y., Jacobs, D. R., and Mozaffarian, D., "Biomarkers of dairy fatty acids and risk of cardiovascular disease in the multi-ethnic study of atherosclerosis," *Journal of the American Heart Association* 2013;2(4): e000092. https://doi.org /10.1161/JAHA.113.000092

INDEX

ABOUT THE AUTHOR

STEVEN R. GUNDRY, MD, is the director of the International Heart and Lung Institute in Palm Springs, California, and the founder and director of the Centers for Restorative Medicine in Palm Springs and Santa Barbara. He is the cofounder of GundryHealth.com, his telemedicine portal for the treatment of autoimmune diseases, IBS, and leaky gut. After a distinguished surgical career as a professor of surgery and pediatrics and chairman of cardiothoracic surgery at Loma Linda University School of Medicine, Dr. Gundry changed his focus to curing modern diseases via dietary changes and supplementation. He is the author of multiple *New York Times* bestsellers, the host of the top-rated *The Dr. Gundry Podcast*, and the cofounder of GundryMD.com, his supplement, skin care, and food company. He has published more than three hundred articles in peer-reviewed journals on using diet and supplements to eliminate heart disease, diabetes, autoimmune disease, intestinal permeability, and multiple other conditions. Dr. Gundry lives with his wife, Penny, and their four dogs (including two new rescues) in Palm Springs and Montecito, California.